For

Congratulations! You have just purchased a great book for someone wanting to build or renovate a home and save money in the process. Why am I confident in saying that this is a great book and probably the best book on the market? Simply because it is easy to follow and understand. It is formatted around my 20 Step Plan which took 14 years, building numerous homes including five homes for my own family to develop. I wanted to simplify my one day seminar for my students so I decided to teach it in the format of a 20 step plan. I found a great improvement, my students had a plan to follow and after just one day they could see the light at the end of the tunnel, realizing that they could easily build their own home following these steps one at a time.

I believe that I can get you up and running with this book faster than any other. In addition to building homes I have worked in banking, and as a Government Housing Officer where I taught night classes on this subject. As well, I've worked in the Apprenticeship Trades Branch and I've completed courses on estimating, appraisal and real estate.

You will learn over 40 money saving ideas! Ideas that I have used to save over $30,000 on four out of five homes that my wife and I built. These ideas could save you a thousand times the cost of this book. You also get a complete management system which includes a forms package for you to use to set up your filing system. You'll want to be well organized! Now I'm sure you'll agree with me, this is a great book and I congratulate you on your purchase and wish you the best of success with your project.

INTRODUCTION

Picture yourself living in a new home in which you have made the decisions about the type of windows, the quality of cabinets, the colour of carpets, and so on. You have also made important planning decisions on energy efficiency, layout of the floorplan, and lot selection. There is no doubt you will take tremendous pride in your decisions and any accomplishments using your own labour. You spend ***your money,*** you make ***your choices***, it's ***your lifestyle***. You get exactly what you want in your home.

Building or renovating your own home is a dream of thousands of people. However, it takes a brave and somewhat ambitious person to come to the decision of turning this dream into a reality. This book will outline a simple step by step plan for you to follow. It will give you sufficient knowledge and the confidence to do this. You can start right now pre-planning the construction by following the (**20 Step plan**) beginning steps which are summarized in the first chapter in order to get you off to a quick start.

The first thing you should understand is that you don't need to know everything about everything to build or renovate a home. You may feel comfortable knowing that before I built my first home, I didn't know what soffits and fascia were[1]. Furthermore, I had next to zero construction experience and I wasn't prepared to take the next fifty years to learn everything about all the different trades. However the secret is to get some key knowledge, the knowledge to take action and make the right decisions, the exact knowledge that I have put together for you in this book.

Technical aspects of construction, such as framing, plumbing, heating, and electrical, are not covered in detail because there are already plenty of books on these subjects. Unless you have the abilities, I don't recommend doing a lot of the skilled trades work. Most important is how to plan, prepare and manage the construction yourself by hiring others who are qualified to do a good job. You will not require previous experience in any of the construction trades. This does not mean the owner-builder can be ignorant when dealing with tradespeople, but it does mean that you do not require previous experience in any trade in order to deal with people and be a good manager. In this regard, the book offers many construction tips and do's and don'ts on the complete art of building your own home.

Contained in this book is **a complete management system.** From a glance at the table of contents you can see that it covers a broad range of subjects. These subjects are all important and I recommend you read the whole book, then use the **20 step plan** and the contents of this book for referencing specific information.

The skills and information needed to build a new home are entirely transferable to renovations. Whether you are building a major addition or simply remodelling your home (replacing carpets, cabinets, painting, etc.) you will find the management tools and money saving ideas offered in this book extremely valuable!

You will find actual job schedules, an estimating checklist, helpful mortgage documents, contract forms and much more. The book is designed to enable you to plan a budget, purchase land, choose an efficient house design, estimate accurately the cost to construct, obtain contractors discounts on everything you buy, apply for a mortgage, get a building permit, qualify for interim financing and strategically organize all of the above into a workable money saving plan. Throughout this book various worksheets are provided for you to use in the book, or you may photocopy them (for your personal use only) to keep in separate files. All of the forms were originally formatted on 8 1/2 x 14 size paper so you can copy (enlarge) them to legal size and enlarge the Recording and Cost Control form (see chapter 13) to the larger 11 x 17 size paper.

The management skills required are within your capability. They become easier to understand as you gain insight from studying this book and take some action. I will show you how to maximize your savings, reduce any risks, and save thousands of dollars! Remember, if you think you can and with my help, you will succeed! Good Luck!
Build it yourself and enjoy!

COPYRIGHT © Regeneration 2000 Inc. tm

David E. Pocock, November 2000
To contact Author www.build-renovate.com

No part of this manual may be reproduced for any means except for worksheets to be used for the purchasers own use. This book is protected by copyright law.

All Rights Reserved.

Cover photo by Anthony Mauza of McMaster Photographers, Calgary Alberta (403) 256-4674

Front elevations illustrated in this book have been supplied by Spectrum Design and Drafting Studios, Box 718, Didsbury, Alberta TOM OWO or www.canadianhomeplans.com

All drawings remain the copyright property of Spectrum Design and Drafting Studios.
To purchase one of these plans contact Spectrum Design directly at (403) 335-8430

[1] See glossary of definitions listed under definitions in the estimating section.

CONTENTS

LIST OF WORKSHEETS

1

PRE-CONSTRUCTION PLANNING

The 20 Step Plan

In this first chapter, I summarized the main planning sections of the entire book for quick reference use and also to allow you to skip over sections which do not apply significantly to your needs. For example, you may already have a building site, house plans or do not have the need for a mortgage or interim financing. You will have an overall view of the work ahead. I advise you to follow these 20 steps in the relative order in which they are numbered. If you don't follow this order then you will likely miss out on some of the money saving ideas presented.

Step 1

While reading this book, which is step one, begin looking for a building site. Purchasing land will come later however for pre-planning your finances you need a ballpark idea of land costs and styles of homes built in that area. Request lot information from land developers. You are only researching availability right now so tell them as little as possible. If you run into difficulty requesting this information, refer to the chapter on purchasing land.

Suppose you are looking for an existing home to remodel, first, make a list of the things you would want (i.e. location, size, price range, style, condition). Next, good deals come and go quickly and you never hear of them because the Real Estate Agents sell them from within their own network of other realtors and clients. You want to be in their network to be advised when something that meets your criteria comes on the market. Therefore, why not network with several realtors to have the advantage of having several others looking for you. Remember, these realtors will not be interested in advising you of property which is for sale by owner so you still have to do some looking on your own.

Step 2

It's best to do a considerable amount of research yourself before approaching a drafting company to draw your blueprint. Even though you should never spend money on plans before finalizing on a lot, nothing is stopping you from gathering ideas and possible alternatives of floor plans.

Some of the elements of a floor plan you can decide on. For example, you may want a walk-in pantry in the kitchen, two bedrooms, a main floor laundry, a den or study and so on. I would look at some house plan books, collect builders profiles or promotional sheets, and start collecting a file of the things we like.

The stationary store at the mall sells drafting skins (large transparent sheets of paper) which I can begin to use to draw out some of my ideas. I would use a regular ruler and a workable scale of one quarter inch equal to one foot. I'm not a draftsman so I won't be going into any detail especially when I haven't finalized on a lot.[1]

Step 3

A whole chapter is allocated to pre-planning construction where step 3 is explained in greater detail. In summary, determine your resources; available cash, qualifying mortgage amount, any work you will do yourself (sweat equity), who you can get help from and your holiday time.

Consider the best time of the year to build and structure your plans accordingly.[1] From your total resources subtract money needed for land and extra items such as a large deck, fireplace, jetted tub, arched windows or any other items which would be considered luxury extras, to arrive at an amount of available money for your base house. You want a rough estimate of just the house with no extras at this time.

This rough estimate can now be used to help you decide whether or not your ideas are affordable and if the maximum size and style of home you want can be built.

Step 4

How much approximately, will it cost per square foot for you to build the size and style that you are considering? This step is to stop you from making the common mistake of committing to a lot or spending money on house plans that are beyond your budget restrictions.

Check other builders prices, ask your realtor and others in the industry to help you. There are a lot of factors that effect cost per square foot which are covered in the house design section of this book. The following are examples of the information you require and are not to be misrepresented as your actual costs.

$60.00/sq. ft. Construction cost for two-storey design
$70.00/sq. ft. Construction cost for bungalow or bi-level
$75.00/sq. ft. Construction cost for split level (including upper two levels only)

Answer the question. Is my plan affordable?

[1] For further details see chapter on house planning and design.

Step 5

Commit to a building lot or a property you want to reconstruct or renovate. Preferably you will have a minimum of six months to one year to get your drawings and to plan prior to construction.

When purchasing land, review a copy of the developers land purchase agreement. Ideally you would attempt to negotiate the most financially advantageous terms first starting with an option to purchase with zero or very little downpayment. Next, a signed agreement with as small a deposit as possible, usually 10 to 15 % of the purchase price.

Some land developers will carry the land interest free for a period of six months to one year. Try to negotiate these terms in your agreement. If you are buying cash, with no carrying terms, ask for something in return, like a discount off the total price. If the developer won't sell to you, refer to some of the ideas in the chapter on purchasing land.

Review your purchase agreement with your Lawyer before signing. Your Lawyer will secure your interest on title.

When purchasing an older property to reconstruct, check with the planning department regarding the rules to hook-up the new water and sewer lines to the city's old water and sewer line. Many cities will require you to pay for replacing the services out to the main line or street. This can cost you an additional six to ten thousand dollars which you must budget for.

Prior to planning an addition, search your title for easements, utility right of way or any other possible restrictions! Talk to the planning department to find the sideyard, front and rear setbacks and maximum lot coverage for your proposed new home.

Step 6

Finalize on a house plan only after you have purchased or placed a deposit on a building lot. The best plans are customized to the sun, slope, shape, all the possible natural advantages your lot may offer.

Take your ideas from your file, (step 2) to an architectural draftsperson, to draw up a rough draft of the main floor plan. Have the plan drafted to a scale you can easily work with, i.e. 1/4" equal to 1 foot. *The same processes apply to building a major addition.*

Your draftsperson should not charge you for a rough[1]. If you are not satisfied with the rough they have created , either he/she will have to start over or else you will have to go elsewhere.

[1] Some drafting firms will charge a nominal fee for a rough which can be applied against the final cost of the blueprints.

Purchasing stock plans from a stock plan company or out of a magazine is an alternative. I have found however, that I always wanted to make enough changes to warrant drafting the entire plan over again.

Construction codes and material specifications will change over time so a purchased plan will have to be checked over carefully, especially if the plan is several years old.

Step 7

Work through your rough floor plan making any required changes. Take the time to carefully think through all rooms in your plan. Some examples of questions you should ask are:

- Is the front door and entrance appealing?
- Is there enough closet space?
- Is the kitchen efficient, open, large enough?
- Is the floor plan functional for our lifestyle?
- Is there too much wasted space?
- What can I modify to make this plan better?

Communicate changes to your draftsperson and continue this process until the final draft is right for you!

Step 8

Have a sketch of the front elevation drawn. Follow the same process, making improvements until the front elevation is exactly what you want. Approve the front elevation before giving the go ahead to draft up the remaining elevations, cross sections, numbering, lettering and other details put on the plans.

Note the following:

1. Curb appeal is critical for future value.
2. Think of the basement layout at this stage.
3. Obtain preliminary approval of your rough plans if building in a controlled area.

Step 9

Refer to the specifications checklist provided and begin to complete this list by checking off those items which apply to you and drawing a line to delete those items which do not apply. You can add to this list if needed.

You will require specifications on nearly everything going into your home to be used for your own estimating purposes. Also, the bank needs to know what grade of materials will be used for appraisal purposes, estimating the value for mortgage approval.

You'll need specifications on type of roofing, exterior finish, windows, carpets, cabinets, mechanical equipment, insulation, interior finishing, fire-place and on framing materials (floor joists, sheathing, studs, beams, flooring). Many specifications will be on the blueprints, others you will need to research by inquiring at lumber yards, truss manufacturers, window manufacturers, etc. and from other suppliers and trades.

Step 10

Blueprints are used for mortgage application, building permit and obtaining multiple tenders from all trades and suppliers. **How many blueprints do you need?**

- **For your own use, 2 sets of plans.**
- **For the mortgage application 2 to 4 sets.**
- **For a building permit 2 sets of plans.**
- **For your trades and suppliers 12 sets.**

Total minimum number of sets of plans - 20

I recommend ordering 20 sets of blueprints plus 10 pages of just the main floor plan. The main floor plan drawings will be used for pricing carpets, cabinets, railings, interior finishing materials and other items only measurable from the main floor plan.

Step 11

Price out your home or renovation by following these four steps in the sequence given.

First, identify (using the cost summary and loan calculation form) what items could be priced by faxing your specifications to various suppliers and trades. Use this method where possible to shop price first, save time and acquire more quotes quickly without incurring any expenses.

Second, locate an estimating service company to complete a material quantity take-off on all your lumber, interior finishing materials including locks and other hardware, insulation and sheets of drywall. This is a small cost and it will give you an unbiased listing which will allow you to compare lumber prices with several companies without having several take-offs, one completed by each company. The take-off formulas in this book can be used if you want to do it yourself. If you have no experience, I do recommend using a qualified estimator to ensure the most accurate material listing.

Third, visit suppliers and trades to get more quotes using your blueprints, copies of the main floor plan and specifications. Request contractor or wholesale pricing. Try to get at least three estimates on every item. Check references and qualifications or certification if you feel it is necessary. Make no commitments at this time.

Fourth, ensure you include all items from the Estimating Checklist listed in the estimating chapter of this book. Summarize all items onto the Cost Summary and Loan Calculation form.

Step 12

Apply for a mortgage! Depending on your needs you will require either a draw mortgage or a completion mortgage.

Prepare your documentation in advance. A list of required documents and some useable forms are provided in the Financing Chapter.

In summary, you will require:

- Personal financial information
- Construction cost information
- House plan information

Follow these simple instructions and you will have the most success!
You already know (from step 3) that you qualify, therefore you should be in control of the interview, i.e. asking all the questions!
You will submit a complete package for the bank to review. (see list of mortgage documents)
You do not need to complete any bank forms until they offer you a mortgage.
I suggest you approach more than one financial institution starting with your own bank first.
Sometimes Mortgage Brokers and Trust Companies are a better source for financing especially if you are close, or slightly over, your maximum qualifying mortgage amount.

Step 13

Apply for interim financing (demand loan or short term line of credit) to bridge between mortgage draws (if draw mortgage) or to complete construction (if completion mortgage).

Request enough funds to complete your entire home. Allow sufficient time for a possible deficiency holdback and lien holdbacks.

Interim financing is considered riskier by the lender because the money is given to you **in advance of construction**, whereas the mortgage draws have the security of the building (either partially or completely finished) before funds are advanced.

Your banker may ask you to spend all your money first which is okay since it will save you some interest expense. Lenders apply these loan criteria for determining approval:

- **Ability** (to make the mortgage payments)
- **Stability** (in your job, business, living address)
- **Security** (in the property appraised value)
- **Age and Health** (based on the size of your request)

Step 14

In most subdivisions today, the land developers or architectural consultants have controls recommending or restricting the colours and exterior finishing materials you can use.

If this step applies to you then contact the developer, ask what information they require and complete their form for approval. I found it is helpful to consult with these experts to see what ideas they would recommend. Obviously, if your neighbours are built and the homes are close together, your selection will be more restrictive.

In most cases you will need this approval before obtaining a copy of the Grade Slip from the developer or the Engineers controlling the subdivision.

Step 15

Apply for a building permit! The information frequently required is the following:

- **Two sets of working drawings**
- **Two copies of a grade slip**
- **Two copies of a site plan or plot plan**
- **Copy of title and any restrictive covenants or caveats**
- **Required fee**

Call the planning department to verify actual requirements and ask about other permits which you may require. Rural areas will require separate permits for a well and septic system. Some cities may require a truss layout or a heat loss statement. If you are building a wood basement, a foundation plan with a professional Engineer's seal will be needed.

For major additions, you will need two sets of plans showing details of the addition and two site plans or surveyor plans of the lot. A development permit may be required if you exceed any existing by-laws with your proposed plans.

Step 16

Arrange for legal, insurance and set up a separate personal checking account to be used for construction only.

Your Lawyer will handle registration of all documents and manage all financial advances from the lender to yourself.

Construction insurance (separate from your homeowners policy) is required by the lender for fire, theft, vandalism and limited liability.

Step 17

Set up your filing system to be organized with the following 12 files. The information for these files can be copied (your personal use only) from the appropriate section of this book. Refer to chapter three for a detailed description of this system.

SET UP THESE FILES

1/ **Mortgage Application Forms** (included in text)
2/ **Trades and Suppliers Directory** (included in text)
3/ **Specifications** (copy working forms in text)

4/ **Cost Summary & Estimating Checklist** (inc. in text)
5/ **Tenders or Estimates** (empty file right now)
6/ **Extra Tenders** (empty file right now)

7/ **Contracts** (copy all usable forms in text)

8/ **Job Schedule and List of Inspections**
(inc, in text)

9/ **Recording and Cost Control Forms** (inc. in text)
10/ **Invoices** (empty file right now)

11/ **Invoices Paid** (empty file right now)

12/ **Documents** (bank, lawyer, developer, city)

These files are necessary to be organized! Sorting through the paperwork is half the battle. Take one hour at a quality print shop, purchase 12 files at a stationary store and you've done this step. I have found it is best to keep all files separate.

Step 18

Finalize negotiations with trades and suppliers. Commit to beginning trades up to the framing (lock-up) stage. Order the windows and roof trusses. You'll be asked to place a deposit when you make these orders.

Set up a line of credit with a major lumber supplier. The local lumber yards will want your business and usually you can negotiate a competitive price with good service.

Meet with your excavator, cribber, concrete company, framer, plumber and electrician. Review the plans, start date, materials to be supplied, some details of construction and any other concerns. Discuss timing and location of services (water, sewer and electrical) with the plumber and electrician. Give a copy of the window rough opening measurements (supplied by the window manufacturer) to the cribber who sets up the basement forms for pouring concrete and a copy to the framer.

For any estimate that lacks the detail required for a good contract, use the contract forms in this book. A copy of these forms should be already in your filing system ready for your use. Just pull out the appropriate contract form prior to your meeting and both fill it out when you negotiate the final terms.

Step 19

Verify Workman's Compensation coverage

for trades working on your home. You are the principle contractor and could be ultimately held responsible in the case of an accident. Each trade must have their own coverage which they pay into. The contract forms also stipulate that they are responsible for this coverage and they carry their own insurance.

Set up personal coverage if you are planning to do any major trade work yourself. You should only pay for insurance when you need it so cancel coverage after completion of your work.

Step 20

Coordinate the first month of activities, which should bring you totally framed up or to the lock-up stage. Start two to four weeks prior to your planned excavation date. Depending on work schedules you may have to book the foundation cribber or the framer several weeks or even a few months in advance.

Arrange for a survey one to two days prior to excavation. The beginning order will be; survey, excavation and trenching for services, delivery of footing material, footing elevation check and replace survey pins for cribbers, soil bearing test if required, water and sewer service, crib footings and pour concrete, crib walls and pour concrete and so on.

See chapter on job scheduling and refer to the forms in your filing system.

This twenty step plan has given you an orderly path to follow. If done in the recommended order which I have presented, this plan should enable you to take advantage of many of the money saving ideas presented in the next chapter.

2

SAVING MONEY

• How I Saved Over $40,000.00 Managing It Myself

Saving money was not the only reason for building my own home but let's face the facts, it would take a lot of years of saving after tax dollars to save that kind of money! If you're planning to be your own general contractor, with one objective of saving money, then here are a few good tips that resulted in big savings.

First, we selected a building lot in an area where the homes ranged in value from $150,000 to well over $500,000. With this wide range our home would not be pegged at one price level. This was important because we were building a 2,700 square foot two storey home and we wanted to maximize the appraised value. Building near very expensive homes helped to raise our appraised value. Location is very important. Our lot is in a lake community and within walking distance to schools.

The next step was to select the right plan. Not only must the plan conform to the building site but it was critical to have a gorgeous curb appeal. We had an expert draftsman customize the plan to the size, shape, location of sunshine, and other natural elements of our building lot.

I must stress how important it is to have an efficient functional floor plan and an eye appealing home. The future value will be greatly reflected in the appearance of the home. The value added benefit of the right lot and the right plan for us was about $10,000.

The third area of savings came from acting as our own general contractor. We did our own research, estimating, negotiating contracts, scheduling and financing. Considering that the preparation time was several months and the construction time four months the resulting savings for doing all this ourselves was about $15,000. If we chose to hire a builder we would have lost control and we would have entered into a long contract spanning more than six months.

The fourth area of savings came from avoiding real estate commissions should we have purchased a house. Though I realize these commissions are paid by the seller, when you purchase a newly constructed home a salesman's commission is usually calculated into the total costs. The savings for us were about $5000.

Fifth, by doing some work myself, painting and some other minor items, I managed to save about $5,000. I could have saved more in this area, however there are trade-offs like having time for holidays, family and interest charges if the job takes longer to complete myself.

The sixth money saving area came from careful selection of products and negotiating with all trades and suppliers. The majority of savings here came from selection of windows, framing materials, roofing, exterior finish, cabinets, interior finishing, plumbing fixtures and floor coverings. The result was additional savings of approximately $8,000. This does not mean I used inferior products. In fact, many products were superior to what most builders in the area were using with the exception of the more expensive estate homes. It does mean that we did a lot more shopping and negotiating better contracts.

Here is a summary of how much money we saved.

Selection of right design and lot	$10,000.
Acted as own contractor	15,000.
Real estate commissions	5,000.
Did some work myself	5,000.
Careful selection and negotiating of materials and trades	8,000.
TOTAL	$43,000.

Total construction costs of home	$185,000.
Professional appraised value (average of two appraisals)	$228,000.
Difference	$43,000.

The savings mentioned above do not include a government new housing rebate which was offering 36% of all goods and services tax (GST) paid. We will receive a rebate from Customs and Excise, a division of Revenue Canada for approximately $3,300.00 (Applicable to Canadian Residents Only)

The following diagram illustrates the potential savings you might achieve building your own home. The land value in this example is one third the cost of the total finished home and land. In some major cities this ratio could be the reverse.

Cost Breakdown of Building A house

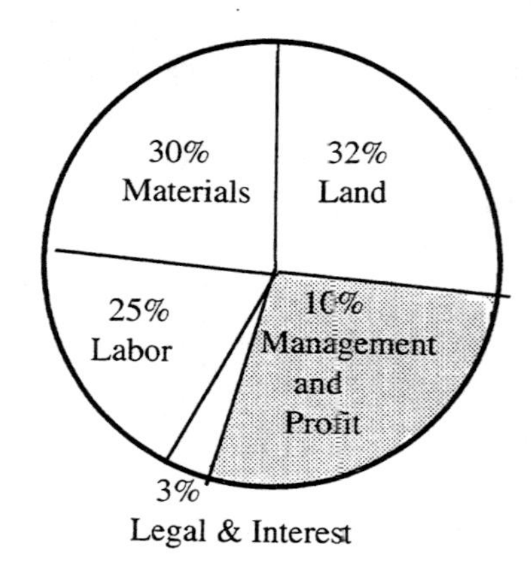

I've presented to you a general outline of how we saved over $40,000 on the last home we built. Now is the time to study a detailed breakdown of all the specific money saving ideas that we incorporated in order to achieve these savings.

45 Money Saving Ideas

Lot Purchase

1 Selection

Since you are custom building only one home, you might be able to find a unique lot not suitable for most builder's designs. Sometimes you can get a good deal on a lot which will require a custom design. We saved $5,000.00 this way on a lot with a utility right -of -way through the back yard. As it turned out, our custom design fit perfectly on the available building space.

2 Lot Purchase

Try to tie up a building lot with an option to purchase. If possible, do not put down any deposit. This gives you some extra time to think about your lot, your plans, financing and budget. When you buy, try to negotiate a better price. Include everything you can think of in your original offer, then ask for something in return if you have to remove some of your original wants. Examples of a want would be a smaller deposit, interest free carrying for six months to one year, include several trees, allow more time to complete the home, etc. *Ask for a few things you don't really need just to give them away in the negotiating process.*

3 Build Closer to the Street

Our neighbor has a runway for a driveway and I know it cost him several thousand dollars more for preparation, concrete and finishing. By building closer to the street you save on the driveway and also the cost to trench and bring in services from the location where they are supplied on your property.

Planning

4 Shop and Compare: Minimum 3 Prices

Simply by shopping around, you will find those trades and suppliers who are willing to offer contractors prices. Many suppliers know you will be shopping around and they could easily lose the business to another supplier who is willing to offer you a contractors price or discount. Rather than have no business, a smart supplier will negotiate with you a price which is substantially less than normal retail. You don't need to set up a business account or even be in your own business, just ask and negotiate. **....Just Ask and Negotiate!**

5 Wait For Specials

Do not commit too early on items that you will not need until 3 or 4 months from now. For example: floor coverings, tiles, deck materials,etc. There are always specials and you may see something better priced later in your travels. Remember to look only for those items which you will likely be supplying such as; light fixtures, Jacuzzi tub, windows, cabinets, floor coverings and all interior wood finishing, etc.

6 Use a Fax Machine for Pricing

You can do up a window schedule and cover letter, and in ten minutes, have faxed your specifications and pertinent information to the window suppliers or manufacturers. Use this method for cabinets, railings, your lumber take-off and any other estimate where you feel it would be appropriate.

7 Don't Get Emotional or Overbuild

A common mistake is to overbuild. I remember spending an extra $3,000.00 on windows and receiving no real additional dollar value when we sold our home Super deluxe windows did not increase the value of our home. When it comes to spending money, determine those criteria that would benefit the future value of your property. Remember there are a lot of trade-offs for the use of your money, including your own retirement savings plan.

8 Negotiate Price AND Everything Else

Always negotiate the price, but also negotiate payment terms, discounts, delivery, options, extra items, etc. Negotiate in good conversation using your best diplomacy. *If you pay peanuts - You get monkeys!*

9 Good Contracts

Most trades will charge extra for items, which were missed on the original estimate and later became your contract. (see framing contract for examples) Ensure that your contract includes all the items you want your trades to do. The contract forms in this book are meant to be used by you for this purpose.

10 Hang on to Your Money

Never give any deposits up front unless it is for a land purchase or as a deposit on windows, trusses, cabinets, which have to be manufactured. Never pay your cribbers, framers, drywallers and other trades a deposit for work before it is complete. If a deposit is required by your trade to buy materials, then you should step in, offer to buy the materials for them, make the check payable to the supplier and ensure delivery to your address. A few simple steps to help secure your money.

11 Assignment of Mortgage Proceeds

Ask your suppliers (concrete, lumber, trusses, windows) if they would accept a legal assignment on the mortgage proceeds to be paid when the home is **complete.** The advantage to them is the guarantee of the assignment (i.e. your Lawyer will make payment after completion and receipt of final mortgage draw) The advantage to you is the 60 to 90 days of interest free credit which will save you some interest financing costs. Sell this idea by explaining that you are a guaranteed sale because your house is sold and the mortgage has already been approved. On the other hand, many builders build homes on speculation through a limited liability company. You are less risk to them than most of these builders.

12 Do Not Buy A Truck

I saved over $40,000. on the last home I built and I didn't even own a 1/2 ton truck! You can have material delivered to the site. If you are short of some material, have it couriered to the site, even if it costs forty or fifty dollars in courier fees. This is far cheaper than buying a truck or trailer. Hire a company to place a garbage container on site and instruct trades to use it. You do not need to haul away garbage. This will cost a few hundred dollars for a complete house, but is well worth it.

13 Lien Holdback

You are legally entitled to hold back 15% of the entire contract when labour and / or both the labour and materials are supplied by your trades. Check the length of time and verify the percentage amount by contacting either the construction association or the appropriate government office.

14 No Bank Fees

You are in control when you go to the bank, if you have completed the steps outlined in the twenty step plan. You qualify for the mortgage and you have all the documentation ready. Tell them you won't pay for an application fee, or a mortgage set up fee and you want interim financing at no more than the bank prime rate plus one percent. You will have to pay for an appraisal fee, but ask your banker to waive all other fees.

15 Use Lots of Blueprints

It will cost you money in time and gas running around town delivering and retrieving blueprints. Use lots of blueprints and don't worry about getting them back. Besides, you may have to make another trip to re-deliver the plans if you decide to use that company. For most houses, a set of blueprints can be run off for between five and eight dollars. Your time is worth more than $5.00.

16 Selling Yourself to Trades

Use the idea; "My house is sold, I have approved financing and I can pay sooner than most builders who want long lines of credit. I,m not building a spec house and I do not have a limited liability company, therefore there is less risk with me".......Negotiate a discount!

17 Buying Plumbing Fixtures

I found it was better to have the plumber supply all the fixtures with the exception of the jacuzzi tub. The plumber buys all fixtures from the wholesaler, some of them like taps, they buy hundreds of one type. It would be difficult for you to get a better deal than the plumber so why not try to get your best deal from the plumber for the complete job. It is much simpler to deal with one source whenever it is feasible.

18 Buying Carpets and Linoleum

Look for those companies that have the inventory on the floor, inventory that they would like to move. We were able to get the linoleum we wanted at a great price, because they had just the right amount left over from another job.

19 Line of Credit

Set up a line of credit with a major lumber supplier, concrete company, wholesale paint supplier and any others you may use. The line of credit will give you 30 days before you have to pay. In some cases, you can extend this line of credit to 45 or even 60 days.

20 Cash Discounts

Always ask for, and take advantage of, cash discounts offered if you pay early. When interest rates are low, a 2% or 3% cash discount will save you money. The cash discount is usually if you pay within 15 days of the invoice date.

21 Government Programs (Rebates)

Take advantage of any government programs. In Canada the GST rebate is 36% of all the GST you pay. Call the Revenue Canada Customs and Excise Tax Department for details and application forms. On one home we built, we were able to apply for a grant for two thirds of the cost of installing a solar hot water heating system.

House Design

22 The Right Draftsperson

Search for a draftsperson with some construction experience which is in addition to producing a good set of working drawings. This person should have a good understanding and appreciation for construction codes, material sizes and proper methods of construction, especially framing. For example; materials come in even size lengths so why plan for odd sizes which will mean more wasted materials.

23 Law of Substitution

If you build a home similar to others in your area, then your home can be substituted for another one when you wish to sell. The real estate agent will compare your home to other like homes that have sold in the area during the past few months. This may pull down or push up the value of your home. In some subdivisions that I have seen, the same box like bi-level design was repeated over and over again, which makes it difficult to sell for more money than what the last similar home sold for.

24 House Type

The lowest cost / sq ft. house to construct will be the two storey design because it needs relatively less foundation, roofing, roof insulation, trusses and excavation. The next house type is the bungalow, followed closely by the bi-level and then the split-level, the most expensive to construct, counting the square footage area of the top two levels only.

25 House Shape

The square box shape costs the least to construct. Whenever you add jogs in the home, you add to wall length and increase construction costs. Even though this is true, it is not to mean that you build a square box. As a matter of fact, the last home we built had a very unique shape giving the home character, defining room space and differentiating our home from the predominate box shaped homes in the area. Consider the best house shape for your lot, the house type, the area and the implications on construction costs and future resale value.

26 Cost Per Square Foot and Appraisal

Try to understand this relationship. All things (quality) being equal, the larger home will have a lower cost per square foot to construct. It might be well worth your while to add on the extra two feet or the den, dining room, or that extra bedroom. Caution! The savings here will be in terms of a possible larger spread between your actual construction costs and the future appraised value. It will still cost more to build the larger home.

As an example, when we built our third home, it was planned to be a 1,572 sq. ft. bungalow. We increased the size of the family room and master bedroom by adding four feet to the rear of the home, bringing the size up to 1,702 sq. ft. The additional square footage cost us about $2,000.00 more but the appraised value went up by over $6,000.00 (representing about $60.00 / sq. ft. at the time) because not all items changed. (see list of items in the "cost summary and loan calculation form")

27 Curb Appeal

The money you save will depend on the future resale value, not your costs. Curb appeal is critical for future resale value. I remember one time the bank executives asking the appraisal firms to do an approximate appraisal by just driving by the property, take a Polaroid snapshot, then from the front elevation only, estimate what they think the value of the property is. This idea was refused by the appraisal institute ,however, it goes to show how much emphasis can be placed on appearances.

I know of many examples of individuals who spent six months to a year of their life planning and building their own home then realizing no financial benefit because their design and ideas were not accepted by the marketplace. It is very important to go through the twenty step plan. Have a rough draft of your front elevation approved before going ahead with the remainder of your drawings.

28 Layout of Floor Plan

Carefully work through your floor plan. View the kitchen as the heart of the home, consider a 1/2 bath or ensuite off the master bedroom and incorporate some vaulted areas in your home. Traffic flow, orientation of windows for sunlight, location of stairs, sufficient closet space and size of foyer are just some of the details which should be carefully considered while designing your layout. A good floor plan will fetch a higher resale value which again saves you money! (refer to chapter on house design)

29 Open Concept

The more open your design, the fewer interior walls. This will save some construction materials. I eliminated some unnecessary walls around the kitchen to open up this area to adjoining rooms.

30 Walk-in Pantry in the Kitchen

Try to design a walk-in pantry in the interior corner of your kitchen. The walk-in pantry is 4' x 4' with an angle on the side facing the kitchen. You can put 5 rows of shelves 16" wide or 40 linear feet of shelving. This is far less costly than extra kitchen cabinets, gives you much more storage space in a needed area and is liked by most people.

31 Staircase

An attractive inexpensive staircase as an alternative to a curved staircase is one which is split 4 rises up, landing, then 10 more rises to the top of the stairs. Use a continuous curve rail and avoid the outdated newel posts. The bottom step should be larger and curved. You can upgrade the spindles and handrail. This can be several thousand dollars less than a curved staircase.

32 Plan for Future Items

When we built our first home, we built the outside fireplace chase, framed in the header and rough opening but left out the fireplace until we could afford to pay by cash. Other items for future development could include the installation of the electronic air cleaner, built -in vacuum system, built- in shelving, decking, finishing the driveway, landsçaping, purchase of a microwave oven and future basement development.

Construction

33 Barter Services for Labor

Do you have relatives or friends in construction with whom you could trade some of your services for work? It's great to get some help but I would advise you not to budget on it just in case your source backs out and now you have to hire out the work.

34 Build in the Shortest Possible Time

Most homes can be built in three months, larger houses 2 to 4,000 square feet can take up to four months, and over 4,000 square feet can take longer than four months. Interest costs on mortgage draws and interim financing will begin to hurt if you take too long to build. Estimating this expense is explained in the financing chapter.

35 Do Some Work Yourself (Sweat Equity)

You can add to your savings of being your own general contractor by doing some of the work yourself. There are jobs that can easily be done with some instruction: insulating, poly vapor barrier, caulking, painting, installing some blocking and backing, cleaning the site, sweeping the house, vacuuming, building a deck and some landscaping. I suggest you leave the major items for qualified tradesmen unless you have the experience and confidence in that trade.

36 Reduce Waste

Save all pieces of lumber 16" or longer. These pieces can be used later for blocking and backing after the framing is done. I found that by using a floor truss system there was less cutting of materials because the floor trusses were designed to run the entire length of the floor. (less cutting - less waste) Just the right amount of material is delivered and every piece can be used with no defects. Stack left over pieces neatly to be returned for credit.

37 Reduce Theft and Vandalism

Buy your locks as soon as the home is closed in; that is, all doors and windows have been installed, including the garage door if you have one. Have 4 keys made and ask your finisher if he/she can make a special trip over to install the locks for you. You will need to loan keys to the trades and hide a key on site. When your lift of lumber is delivered, check the packing slip and then mark the ends of the lumber with spray paint for identification and to deter theft.

38 Save Money on Energy Efficiency Products

The best energy efficient idea for your money is the thin inexpensive poly vapor barrier. It is sometimes referred to as the critical vapor barrier because the insulation has no value if the cold air can just blow through! Use 6 ml. poly but before you install make certain the insulation has no holes or cracks and every joint between framing top and bottom plates, corners, headers, beams, etc., is caulked to help seal off air infiltration. I also wrap my windows and doors with poly to overlap with the poly on the inside. The two are sealed together to form an airtight seal. *(use an 18" roll and hammer stapler and seal to the window frame with caulking leaving an extra six inches on the corners)*

39 Wood Basements

If you are considering a bi-level or raised bungalow, a wood basement may cost less than concrete. Since your foundation walls will only be 4 to 5 feet into the ground, the specifications for the wood basement will call for less material than if your basement were 6 feet into the ground. The wood basement will be cheaper to finish afterwards because there is no additional framing required. Wood basements are generally warmer and dryer than concrete. They are especially good in rural areas where it becomes expensive to haul concrete long distances.

40 Excavation and Trenching

When you excavate, see if you can have both the excavation and trenching done at the same time. Sometimes the same equipment can do both jobs which could save costs in equipment transport and you may get the trenching included in the cost of the excavation. Also I have found savings by lumping all excavation, trenching, backfill, grading and loaming into one contract with one reputable company.

41 Footing Forms

Check with the basement cribber, they may have their own form material which can save you this expense. When I do buy footing forms, I like to use 2 x 6 instead of 1 x 8 boards. The 2 x 6 can later be used for bracing when the concrete walls are being poured and afterwards used in the framing stage.

42 Extra Basement Height

Most basements are a standard 8 feet in height. Some builders will go the expense of a 9 foot high basement to allow extra room for ductwork which will give a full 8 feet or more of headroom. As an alternative to this expense, use a 2 x 6 ladder for cribbing the top of the concrete wall instead of a 2 x 4 ladder. Ask you cribber to give you the additional 11/2 inches in height. You can cap the top of the concrete wall with a 2 x 8 plate to give you another 11/2 inches. The 3" of extra height may be the extra height you need for your future basement development. By using floor trusses and a little planning you can hide most if not all your heating ductwork and plumbing drains.

43 Basement Beam

I eliminated an expensive 4-ply, 2 x 10 beam by putting in a bearing wall. This was convenient, as we were going to have a wall there anyway. The net savings in beam material and teleposts was approximately $500.00. The whole idea here is to look for trade-offs and alternatives for every item going into your home, selecting those that can save you money without taking from the total future value of the home.

44 Fireplaces

Gas fireplaces can be directly vented out the wall, eliminating the need for a complete fireplace chase. You will require a "doghouse" or small chase extending the height of one room. The savings will be anywhere from $500.00 to several thousand if you were considering a brick fireplace chase.

45 Interior Finishing (baseboards, window and door casings)

You can create a nice effect with a combination of two finishes in your home. For example; with the use of an oak finish for your cabinets, baseboards and trim around the doors, the second finish could be paint grade window trim, using less expensive pine or aspen. Note that the softer wood is used only around the windows. The more expensive hardwood is used in those areas that are exposed to getting bumped and banged and will still look good after many years of wear and tear.

The strategy you should follow to maximize your potential savings is obviously to use as many money saving ideas as possible. Remember the amount you save should only be measured against the future worth of the property. For example; suppose you can save $20,000. by building yourself as compared to a quote from a builder. It's possible, to still be in a negative equity situation if you build an ugly design, overbuild, build in the wrong area, build a box similar to all the other homes, etc.

These money saving tips have helped me save over $40,000.00 on just one single family home. You will become more familiar with many of these ideas as you follow through with the remaining chapters. In the meantime, you should begin to apply these ideas whenever possible so you too can begin to **save money!**

3

MANAGEMENT

You As A Manager

In this chapter I will talk to you about being a general contractor and how to set up your management system. The purpose of this section is make you aware that the dominate skills required to build your own home are not framing, plumbing, electrical, etc., but management. Too many people have failed or run into serious trouble because of a lack of management knowledge.

Building your new home may require you to take out a large mortgage and use all your savings, which can create a strain on you and your family. You need to be mentally prepared for the job.

When you make the decision to build your own home you are, for all practical purposes, starting your own small business. You will be involved in purchasing, signing contracts, scheduling, and accounting with the end objective of saving or making money. Since you are performing as though you are operating a small business, you must also conduct your efforts in a professional manner as if you were a business manager. To be successful you must place yourself in a manager's frame of mind.

We will look at your functions as a manager relating to contracting and promote you to the position of president. If you are building your own home, the ultimate responsibility for every tradesperson, supplier and banker is on your shoulders. Starting now, view your project as a challenge - an opportunity to learn and to improve your lifestyle.

The manager's job is to guide the work to a predetermined end within the limits of a budget. The management functions are very similar to many businesses, from a small office to a large factory. The type and volume of work differs but it is the management skill that is essential. Study the following important functions of management and their application to general contracting.

Planning

Planning is decision making, that is, deciding what to do, how to do it, and when to do it. It is essential for your success. It should take you as long or longer to plan for construction as it will to actually build the house. Where do you start?

You have to have goals. Every successful manager will tell you to set goals. Once you establish these goals or objectives they act as a motivating force - a plan for action!

Life is more exciting when you have goals to work toward. You will have to make a commitment sometime and take some initial steps if you want your goals and dreams to come true. You took the first step by buying this book. The second major purchase is a lot to build on. But first, you begin by setting some goals and planning the beginning steps to achieve those goals.

Organizing

A big part of your job is to get yourself organized. In the latter part of this chapter you will learn what information should be included in your management system and how to set this system up. This was summarized for you in the 20 step plan.

You may not be aware of it, but smaller sole proprietorship and small business, on the average, are more profitable than large multinational corporations. Why? One manager can be far more efficient and effective in control of his or her own resources than a group of managers in a large business which has more overhead and more waste.

The larger builder has the advantage of mass production and volume buying, thus keeping costs lower. You have no overhead, no salesmen or real estate commissions to pay, no salaries to cover, you can do some of the labour yourself and your biggest advantage, which can put you ahead of the builder, is that you now also have an excellent management system which organizes your whole project from start to finish. Stay organized by using this system!

Hiring

As a general contractor, your key to successful building and savings will be in your research to find competent subtrades. You can only do as good a job as those who work for you.

Plenty of research will locate trades with the ability to do great work at the right price. After you find them and have checked out their qualifications and certification (if required), then the next step is to negotiate a contract where both sides know exactly what is to be expected from them. If this is done properly at this stage then you will have eliminated nearly all chances of having trade related problems later on. Just imagine the problems you could run into if one vital piece of information was missing from a contract, for example; contract does not indicate "time required to complete the job".

Directing

After starting construction your job is to develop the right rapport to bring great work out of your tradesmen. Every manager needs to keep track of employees' performance. The most pleasant way is to have influence with people rather than to exert power over them. The same is true for the people you contract with.

There are many ways of issuing orders, but only one of them is the best way in a given situation. You need to determine which way is best. Most of the time, just asking pleasantly will get the right result. A wise manager is tactful, uses diplomacy, and assertive but never unpleasant. Harsh, direct orders usually create dissatisfaction and result in lost time and poor workmanship. Giving orders may be the toughest part of your job as construction manager if you have never been in such a position before.

Your function of directing involves **communication** of ideas, suggestions and objectives. This includes asking plenty of questions and listening to the suggestions and complaints of others. Every worker is entitled to a clear-cut understanding of what is expected. This seems to be an elementary requirement, but it is surprising how many efforts have failed or have been botched by ignoring it.

Controlling

There are several forms of control you have working to your advantage to ensure quality is maintained. First, you have specifications on drawings by a qualified draftsperson or architect detailing information such as the sizes and types of lumber, strength and reinforcing of concrete, grade of shingles, location and number of electrical plugs and many other details of structural requirements.

Next, you have the starting specifications included in this book which will soon be in your filing system ready for your use.

Third, you have the building codes that all trades must follow when building your home. However, you must still check certain items during construction. For this purpose I have provided you with a checklist of items to review with each trade during the construction. I recommend visiting the job site daily to check on the progress and discuss any work or scheduling issues with those on site. If necessary, follow this up afterwards, in the evening, with a phone call to the person you contracted with. Keep those lines of communication open and active!

The fourth area of control is through the required inspections. In most cases, depending on your mortgage and location, you will have to have inspections at various stages of construction. These inspections are required by the town or city, the mortgage lender (for financing purposes), and the provincial (state) inspectors for plumbing, electrical and gas. These inspectors are really working for your best interest to ensure all codes and safety measures are met and government regulations are being followed.

In addition to construction, a general contractor must also control the finances by means of a budget. This involves follow-up on all advances, interest expenses, invoices, and most important, it means having the mental control to spend only within your means. The forms in the management system to record your expenditures will help you control spending and stay on budget.

Co-ordinating

Co-ordinating involves proper scheduling of materials and trades so everything works in harmony. Each subtrade has a series of duties to perform. Your management function is to respect their timetable by co-ordinating it with your own construction timing schedule. The secret is to maintain two-way communication on a daily basis with all those involved in your construction schedule.

Some important information to obtain from your trades is the lead time required to start the job and also the length of time required to complete the job. Both of these items are included in the contract forms provided. In the job scheduling chapter you will find out how all this fits into a critical path and your own job schedule. Your goal as a co-ordinator should be to complete the project in the least amount of time, reducing interest costs and potential delays.

Innovating

Building your own home, the very nature of it, requires you to be somewhat innovative. This may involve developing new ideas, or combining several already proven ideas, tackling the unknown, doing things in new ways and offering alternative solutions to problems.

Your house design is where you are most likely to be innovative. Your plan will be different from others on the block and will be designed to suit your family's needs. Right from this stage, the whole process requires you to make hundreds of decisions. Now may be your only opportunity to incorporate all those marvellous ideas into your plans.

Motivating

To be a successful builder requires leadership and initiative to get the operation off the ground and carry it through sometimes in spite of discouragement. This demands thinking positively, motivating yourself and others.

You will be successful by drawing the best out of your trades and suppliers. Treat them as though they are your business partners and your business depends on their quality workmanship. Be optimistic by looking for solutions to any problems rather than looking for someone to blame. A business manager who has a problem will first look at all the alternative solutions and discuss the situation with all those affected, rather than look for someone to blame. Then he/she will analyse each one according to advantages, disadvantages, strengths and weaknesses, then choose the alternative that offers the most compatible solution, communicating the outcome to those affected.

Motivating is a very important function of your job as a manager. Your success will be determined by your mental attitude. Good tradesmen are often very busy. I have known a few tradesmen to charge extra on the original quote just because the builders attitude towards them was not as pleasant as they would have preferred.

Motivation starts with you first! If you do not have any enthusiasm yourself, you will not succeed in persuading people to work enthusiastically.

Representing

As the builder, you will be the representative in charge, responsible for all construction and any disputes with trades or suppliers. You are also your own representative in your dealings with the lawyer, insurance agent, banker, land developer, engineer, architect consultants and all inspectors. Each has responsibility to you and to others. Similarly, you have responsibilities to them, especially to be informative.

At times you may feel that there is far too much red tape and that everyone is against you, but it is important to take a positive approach. You are fortunate to have this management system and if you will just ask others for advice when needed you will see that you have much experiĕnce available in various fields willing to give you assistance.

How To Set Up The Management System

The worksheets discussed here are designed to help you set up an efficient management system which will save time and money. To quickly set up your system, follow these instructions:

1/ Refer to the Table of Contents to locate the worksheets you want to copy.
2/ Photocopy all the worksheets and contracts making multiple copies where required.
3/ Organize into twelve legal size file folders.

Mortgage Application Forms and Worksheets

The documents you will require for mortgage financing are listed in the worksheet titled **CHECKLIST FOR MORTGAGE APPLICATION.** Copy all the worksheets and forms from this chapter. In addition, from the financing chapter, copy the **MORTGAGE CALCULATION WORKSHEET, MORTGAGE PAYMENT TABLES** and worksheet titled **ROUGH ESTIMATE FOR THE COST OF BUILDING.** These pages must be completed now to ensure you qualify for the financing in advance of purchasing land, blueprints, etc. Add to this file the other information required for your submission (see checklist) which will be obtained after you purchase land, have plans drawn up, site plan drawn, etc.

Trades and Suppliers Directory

Place in this file a copy of the blank form **INFORMATION DIRECTORY and TRADES AND SUPPLIERS DIRECTORY** chapter 4-8 to 4-10. The blank forms are for you to complete just prior to construction. You can start completing the information directory right now. These forms will be your communication directory and you will refer to them often for phoning and scheduling work.

Specifications

The specifications checklist can be put into a file titled **SPECIFICATIONS**. Use these pages as your starting list to customize your own specifications. Simply tick off the items you choose for your new home or renovation and delete the items you don't want. This will be a working list to use while you're pricing out your home. You may add to this list some custom items of your choosing. The specifications will be used later as one of your mortgage documents to determine the appraised value of the home for mortgage financing purposes.

Cost Summary and Estimating Checklist

The **COST SUMMARY AND LOAN CALCULATION FORM** (Cost Summary for short) is very important because it summarizes all your estimating onto one very functional form. This form will eventually become one of your documents for mortgage application. As you can see, you should have the entire form complete prior to applying for a mortgage.

I suggest making three or four copies of this form to have on hand because you will be adjusting the figures frequently until you begin construction.

The 28 numbered items on this form summarizes the **ESTIMATING CHECKLIST.** Copy these pages and staple behind the Cost Summary and place in a file titled **COST SUMMARY AND ESTIMATING CHECKLIST.** (Refer to the Estimating Chapter for learning how to use the Estimating Checklist.)

Tenders (or Estimates)

This is simply an empty file right now. It should be used for all the estimates and product information you receive from all the suppliers and trades. Put in this file only those estimates that you have selected to be the best and you plan to make a commitment on. Organize all estimates according to the order shown on the Cost Summary and Loan Calculation Form.

Extra Tenders

Set up a separate file for those **EXTRA ESTIMATES OR TENDERS** that you do not plan to use.

Don't lose them or destroy them because you just may need to go back and use one of these companies should something happen to your first choice.

Contracts

Place in this file one copy of all the contract forms in the text. These contract forms can be used after you have done all your pricing and have selected your trades and suppliers. No need to provide **CONTRACTS** when you are only doing your initial pricing! Use these documents afterwards to negotiate your contract, terms and conditions with your trades. Many times I have used a well prepared estimate from a reputable contractor as the basis for the contract however, it is important to ensure that all relevant information is listed or added to their estimate.

It is important to have one copy of each **CONTRACT** form on file so it is available when you need it without having to race around to get copies made or forgetting to do it altogether.

Job Schedule and Inspections

Several pieces of information, reference materials and charts, should be included in this file for your quick reference. The **JOB SCHEDULE AND INSPECTIONS** file should contain the following:

- **MASTER CONSTRUCTION SCHEDULE**
- **SAMPLE JOB SCHEDULE**
- **SAMPLE BAR CHART**
- **LIST OF INSPECTIONS**
- **CALENDAR**

All the above listed forms with the exception of the calendar are supplied in this text.

Every owner builder's construction schedule will be different; trades, contracts, lead times, times to complete each job, different size of each home and so on. The information in this file is to enable you to plan and complete the master construction schedule which will custom suit your project. (see chapter on job scheduling for full instructions)

Recording and Cost Control

The **RECORDING AND COST CONTROL** form is so you will know exactly where you are with your financing at any time during construction. Record every financial transaction on this form when you write cheques, make deposits, balance your checkbook and compare estimated to actual costs. Each item listed under the "description of item" column will be numbered corresponding to the numbers from 1 to 28 located on the Cost Summary and Loan Calculation form. In this sequence, the **RECORDING AND COST CONTROL** form is tied directly to the Estimating Checklist. When you photocopy this form you should enlarge it onto 11 x 17 paper and make 6 copies.

Invoices

Set up a file titled **INVOICES** to initially hold all the invoices and statements you will receive from all the trades and suppliers. This file will hold invoices for a number of days, allowing you to take advantage of any lines of credit, until you do your payables, when you will write and record your cheques.

Invoices Paid

The purpose of this file is to allow you to sort between invoices received and invoices paid. It is simply a storage file for all your paid invoices.

Documents

The **DOCUMENTS** file is simply another storage file for many bits of information and letters you will receive from the inspectors, your lawyer, bank, land developer, engineers, architectural consultants, etc. This file should contain the copies of your grade slip, site plan, inspection reports, certificate of title, and other legal and bank **DOCUMENTS.**

Aside from your management system, you should have a personal file for your mortgage and approval letter as well as a separate file for your property insurance.

HOW DOES THE MANAGEMENT SYSTEM FUNCTION ?

- The specifications and cost summary forms will be used for part of your mortgage application.
- The trades and suppliers directory is for communication purposes used during job scheduling.
- The cost summary is tied into the recording and cost control form using the same numbered listing for ease in cross referencing.
- Tenders will eventually become your contracts or use the contract forms included with this system.
- Some contract information is used to complete the job schedule.
- All invoices are paid and recorded on the recording and cost control form.

(see diagram)

You have come a long way in a very short time. You can see from just the first three chapters that it is important to organize yourself so that the information is simplified and quick and easy to refer to. Now is a good time to set up your management system following these instructions or you can wait until you reach this step as explained in the 20 step plan covered in chapter one. The remaining chapters of this book will provide you with valuable information and all the forms supporting what you already learned.

HOW DOES THE MANAGEMENT SYSTEM FUNCTION ?

FLOW OF INFORMATION

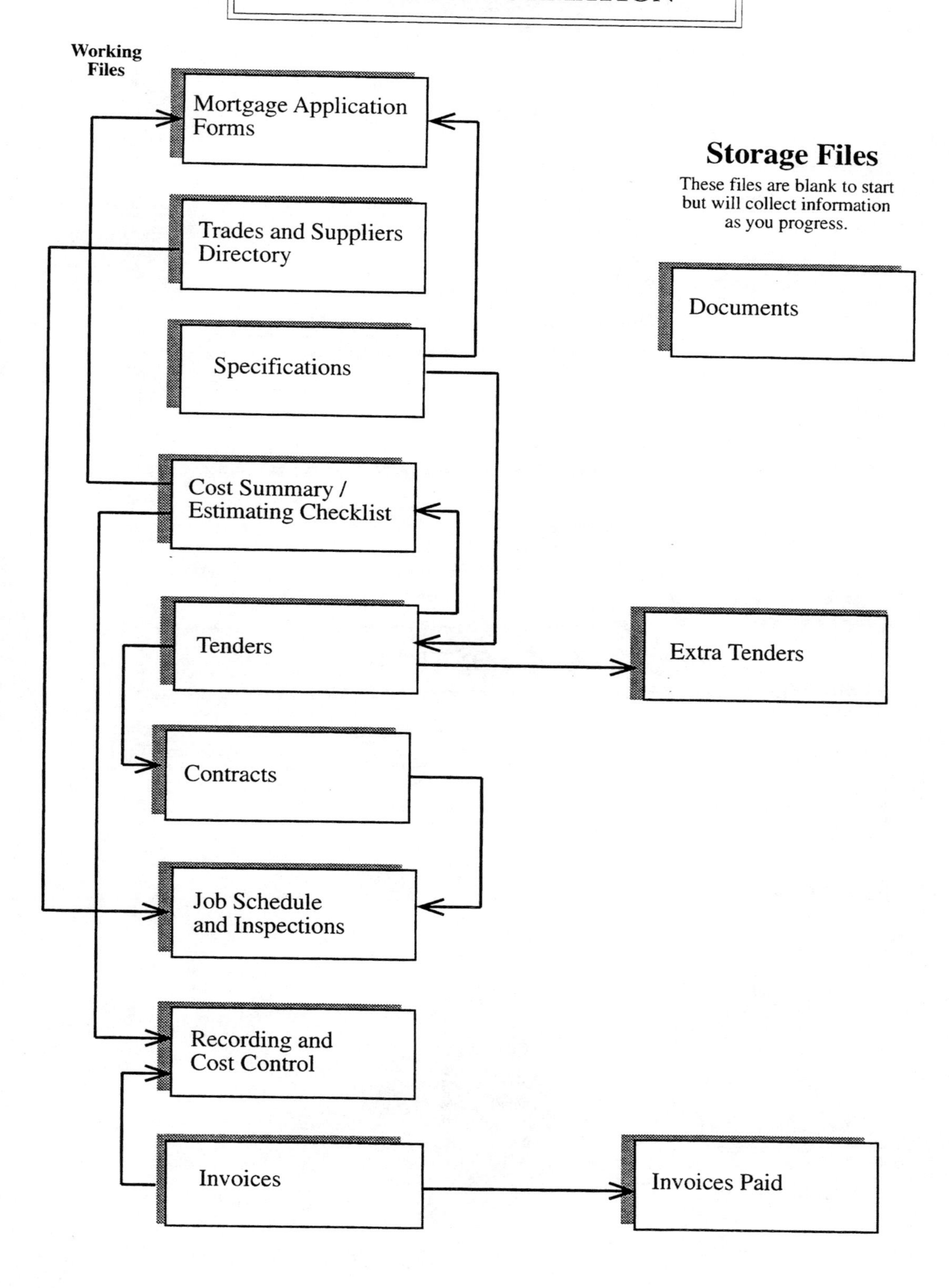

LARGE LABELS FOR 8 1/2" X 14" FILE

MORTGAGE APPLICATION FORMS 1) Personal Financial 2) Cost Summary 3) House Plan forms and specifications
TRADES AND SUPPLIERS DIRECTORY
SPECIFICATIONS
COST SUMMARY & ESTIMATING CHECKLIST
TENDERS OR ESTIMATES
EXTRA TENDERS
CONTRACTS (Samples from book)
JOB SCHEDULE AND LIST OF INSPECTIONS
RECORDING AND COST CONTROL
INVOICES
INVOICES PAID
DOCUMENTS Bank, Lawyer, Developer, City Inspections, etc.

4

PLANNING YOUR STRATEGY

Planning Your Strategy

How do you begin to implement an idea? Start by appraising present and potential opportunities, resources, strengths and weaknesses. Second, carefully consider the alternatives, the advantages and disadvantages and potential cost of each. Third, decide which alternatives are best to achieve your end objective. Finally, begin carrying out your plans.

I can remember planning for our third home. We wanted to down size to a smaller home for financial reasons. We had no money for a lot or blueprints. How did we do it? Well, we took out a $15,000. second mortgage on our existing property to cover the deposit on a lot, the drafting of house plans and the cost of a building permit. The terms on the land agreement allowed us one year to pay the balance. We priced out our planned new home and selected all our trades and suppliers. The new construction or draw mortgage was conditional upon the sale of our existing home. Our home was listed and sold with a ninety day possession. We started construction the day after our house sold and completed one week before the end of the ninety days. Pre-planning really paid off, we only moved once and construction all went smoothly.

Determining Resources

Guard against allowing your planning to wonder off into a dream world. Planning is an exceedingly practical job. Be realistic; objective honesty is essential. Write down the answers to the following questions.

- *How much money do I currently have to invest?*_________________
- *What amount of mortgage do I qualify for?* _________________
- *How much money can I save prior to construction?* _______________

Answer these questions and you will soon be able to determine what you can afford to build. Use this simple formula:

Maximum total construction costs=Your present cash input + Your qualifying mortgage amount.

Next, decide what resources you have available to help you either save on construction costs or increase your cash input. The answers to these questions may help.

- *Who can I obtain help from?* _____________________________
- *What work can I do Myself?* _____________________________

- *How much time do I have for my holidays?* ____________________
- *What can I sell to increase my cash input?* ____________________

Plan to do only the work you are capable of doing and within a reasonable time span. Taking on a major project like framing, roofing, or interior finishing may not save you any money if it takes you twice as long to complete. You have to pay for accumulating interest costs. Be confident that you have the ability and time to do a job properly and quickly. If you don't, contract out these tasks.

Provide for contingencies and ensure that you have enough money in your budget to contract out everything, just in case you don't have the time or cannot do it yourself. This helps to avoid having to obtain additional financing.

Evaluating Alternatives

Once you know what resources you have to work with, you can evaluate present and potential opportunities. Begin by researching some of these possibilities.
Ask:

- *What size house should I build?* ____________________
- *What type of house should I build?* ____________________
- *Where should I build?* ____________________
- *How long do I intend to live there?* ____________________
- *How strong is the resale market and will my plan be saleable?*
____

Always consider the resale market when planning your design and location no matter how long you intend to live there. If you have to sell, you will want to get a good dollar from your investment. If you intend to live in the home for only a couple of years, you would build not to suit your own needs but the expected needs of the real estate market. In a poor resale market, you may decide not to build at all unless you can do a fantastic job of planning and build for less than what similar homes are selling for.

In order to properly assess your alternatives, you need information regarding average cost per square foot to construct and some insight into how a home is appraised and what things influence value. You can start now by collecting information on alternative house designs, financing, land costs, and the values of homes in various areas where you can build new or renovate. Planning of this nature will give you the answers to these questions.

- *Approximately how much will the house cost?*____________________
- *What will the mortgage payments be?*____________________
- *How much will the house be worth?*____________________
- *Is it better to build new or buy and renovate?*____________________

The house design you choose has to be affordable. Your research is perhaps the most critical step of all. Here, you should spare no expense in your quest for facts. The entire process may take several months of your spare time. Building a home may be one of the biggest decisions of your life, and you don't want to make any errors that may be costly later on. Share your plans with as many people as possible: realtors, friends, relatives, etc., and have an open mind to their thoughts and ideas.

Implementation

Your next step is to begin developing a pre-construction plan that incorporates the timing of your decisions. This plan will give you definite goals. Begin by writing down answers to the following questions.

- *When should I build?* ______________________________
- *Do I sell my existing home first?* ______________________
- *Should I place a deposit to secure a lot now?* _____________
- *How long will it take to build?* _______________________

The objective is to suit your needs, reduce construction time, minimize interest costs, reduce risk, and maximize your savings!

If you already own a home now, but intend to build a new one, study the alternatives on the following page and note the advantages and disadvantages.
Avoid alternative number one unless you are wealthy. If you are a reasonable risk taker and the market is good, you might consider alternative number two. If you want to be safe, choose number three. You will have to move twice, however, for many it will be the wisest decision.

Review and Revise

Continually study your plan for new opportunities to economize. Plans seldom remain fixed for all times. As with business, goals are achieved or abandoned. Incomes increase or decrease. Market conditions move up or down. Make sure you can adjust your plans accordingly to reflect any changes.

Pre-Construction Critical Path

The pre-construction critical path illustrates the series of steps you have to take (20 step plan) to get to the point where you are able to begin construction. The word critical applies to those functions that cannot be performed until something else has been completed first. Follow this path to preplan your construction in the shortest time frame. For example, setting up your management system was shown as step # 17 in chapter one. As illustrated in the diagram, this step is not in the critical path therefore it can be completed at any time prior to construction. (see Pre-Construction Critical Path)

Pre-Construction Schedule

Once you have made your plans and decided on the procedures you will follow, set definite goals by writing down the dates you will complete each step. Use the pre-construction schedule provided.

PRE-CONSTRUCTION CRITICAL PATH
(Based on the 20 step plan)

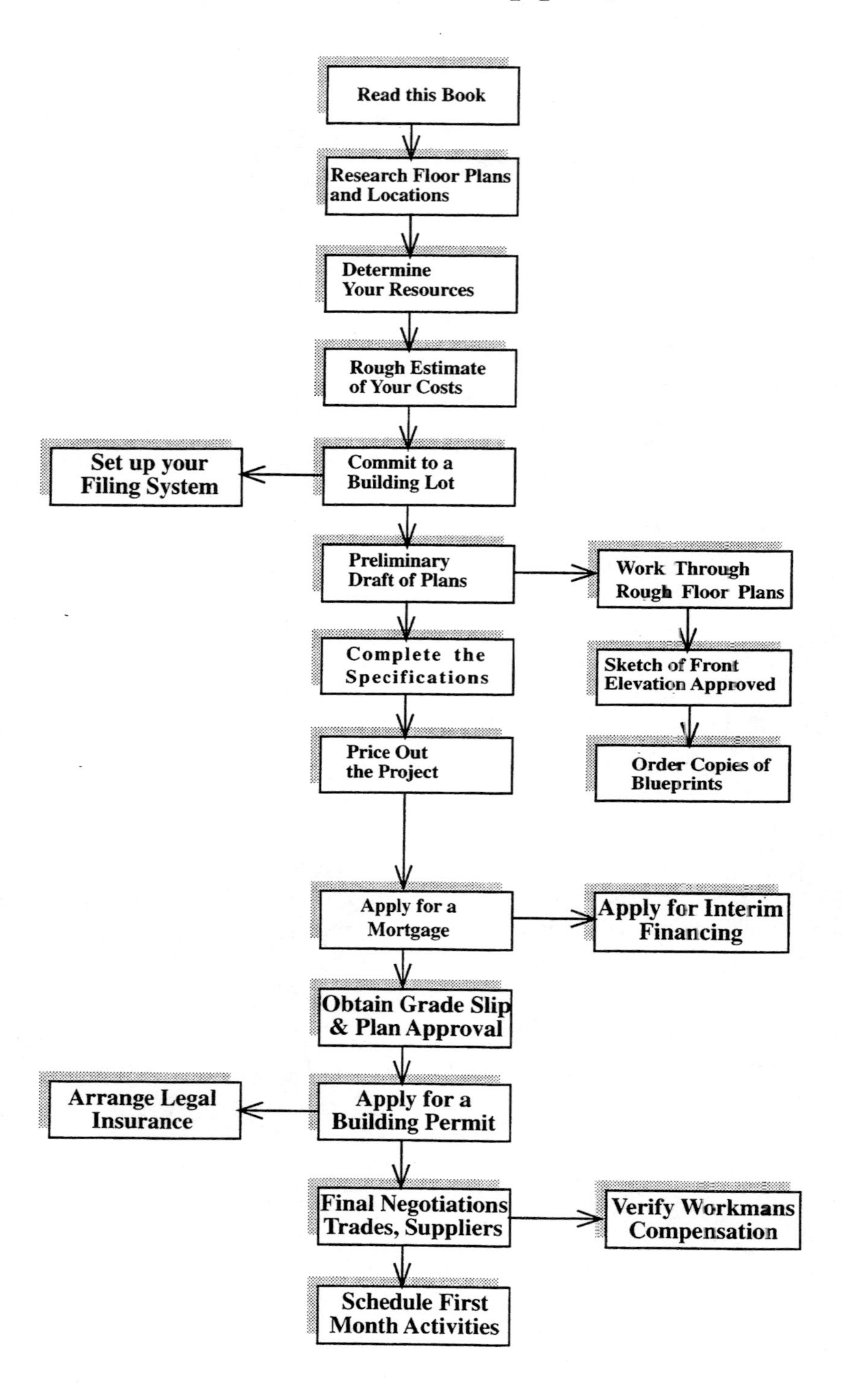

Three Possible Alternatives

Alternative	Advantage	Disadvantage
Begin building new home prior to selling current one.	Can move directly into new home without moving twice.	High risk that present home may not sell. May have to list both homes or rent one to avoid paying carrying costs of two homes. May have to sell below your expected or appraised value.
Purchase lot now (by placing a small deposit), begin planning, costing, apply for a building permit but do not start until existing home sells.	Risk is fairly low in a good market; however, the market can quickly change. Possible to begin construction as soon as home sells. May be able to move directly into new home if 3 to 4 months possession is obtained. May be able to get a good deal on property now that may not be available later.	Accruing interest on lot may make it unaffordable. Building commitment may require you to pay out lot in one year, which may be too costly. Land would have to be increasing in value to induce you to buy now.
Wait for present home to sell prior to investing in property and building.	Low risk because you do not have to spend or borrow money. Can wait for a good offer on current home. Monthly payments kept to a minimum.	Little incentive to take action due to non-commitment of funds. Cannot plan and set goals until existing home sells. Will have to move and rent then build and move again.

HERE IS MY PLAN FOR ACTION (INCLUDE DATES)

Worksheet # 1

PRE-CONSTRUCTION SCHEDULE

DATE	ACTIVITY	COST

Tips On Being Organized

- Plan ahead.
- Purchase a business card file folder.
- Set up your filing system as explained in chapter three.
- Complete the information directory in this chapter.
- Maintain a diary of all communication between subtrades and suppliers.
- Always have a note pad and pen available.
- Start recording all addresses you will need to send your future change of address to.
- Plan your holidays carefully; save time for moving.
- Follow a schedule.
- Obtain plenty of copies of your house design.
- Plan to do only the work you are capable of doing yourself.
- Visit the local Government Housing Departments to pick up any available booklets on house construction.
- Obtain information on current building codes and regulations.

If you are well organized you will be able to plan and build your own home in your spare time.

Worksheet # 2
INFORMATION DIRECTORY

	Contact Person	Phone	Address
Mortgage Company Lender Inspector			
Municipality or City Zoning Building Permit Inspections			
Land Developer Architectural Consultants Engineers Surveyers			
Utilities Gas Company			
Power Company			
Telephone			
Cable T.V.			
Other			
Insurance			
Architect or Designer			
Builder or Project Manager			
Other			

TRADES AND SUPPLIERS

PROJECT	COMPANY	CONTACT PERSON	PHONE NO.	PRICE	% OF TOTAL
1. BLUEPRINTS					
SURVEY					
INSURANCE					
2. EXCAVATION, ETC					
3. BASEM. CONCRETE					
WEEPING TILE,					
DAMPROOFING					
GRAVEL					
4. CRIBBING					
5. BASEMENT FLOOR					
CONCRETE					
GRAVEL					
6. FRAMING MATERIALS					
TRUSSES					
STAIRS (3)					
7. FRAMING LABOUR					
8. DOORS & WINDOWS					
GARAGE DOOR					
SKYLIGHTS (2)					
9. ROOFING					
10. STUCCO					
BRICK					
11. FASCIA, SOFFIT,					
EAVES					
12. PLUMBING CONTR.					
MARBLE, TUB					
13. HEATING					
14. ELECTRICAL					
INTERCOM					
15. INSULATION					
16. DRYWALL					
BASEM.INSULAT.					
ATTIC INSULAT.					
17. KITCHEN CABINETS					
18. FLOOR COVERINGS					
19. TILES, FOYER,					
BATHS					
TILING LABOUR					
20. ELECTRICAL					
FIXTURES					
21. PAINTING (MAT'L)					
PAINTING (LAB.)					
STAIN & LABOR					

TRADES AND SUPPLIERS

PROJECT	COMPANY	CONTACT PERSON	PHONE NO.	PRICE	% OF TOTAL
22. INTERIOR FINISHING					
MATERIAL					
LABOR					
RAILINGS					
MIRROR DOORS &					
MIRRORS					
23. PARGING					
24. FRONT STEPS					
WALK,DRIVEWAY					
GARAGE PAD,					
GRAVEL					
25. EXTRAS					
26. SERVICES/PERMITS					
GAS LINE HOOKUP					
CABLE TV					
A.G.T.					
GARBAGE					
REMOVAL					
27. BUILDERS COST					
28. LEGAL FEES					
INTEREST ON LAND					
MORTGAGE DRAWS					
DEMAND LOAN INT.					
BANK					
TOTAL CONSTRUCTION					
COSTS					
BEAMS & PRESERVED					
WOOD FOUNDATION					
MATERIAL QUANTITY TAKE OFF					
MORTGAGE CONSULTANT (BROKER)					

5

PLANNING A BUDGET

This chapter deals only with preliminary budgeting that should be done prior to purchasing land and investing in house plans. The chapter on financing, on the other hand, deals with your mortgage application, mortgage documents and interim financing.

I recommend completing the worksheets in this chapter before investing in land or spending money on house plans. You will learn how to calculate your maximum qualifying mortgage amount and how to do a rough estimate of the cost of your proposed home prior to spending a dime of your own money.

These two calculations can be critical for many people who are on a tight budget and likely building to maximum resource capabilities. It can help you to do a rough calculation showing you what you can afford, however, please don't confuse this with the job of estimating which gets into every component of the project.

Calculating Your Maximum Qualifying Mortgage

You can easily calculate the amount of mortgage you qualify for. The first step is to check with your possible lenders for their current interest rates and their gross debt service ratio (GDSR). Most lenders allow about 30% of your gross monthly income to be applied towards mortgage payments and taxes, often referred to as your principal, interest and taxes or (P.I.T.) for short. With some lenders the percentage used for calculating the (GDSR) may vary 2 or 3% and others might include an estimate of the heating bill with the mortgage payments. However, for the most part, and for the examples I show, the ratio used is 30%.

A second important ratio you need to be familiar with is the total debt service ratio or (TDSR) for short. This ratio refers to the maximum an applicant is allowed for PIT plus all other debt obligations. For example, a yearly income of $48,000, or $4,000 per month, with a 40% TDSR would mean maximum total debt payments of 4,000 x .4 or $1,600. The GDSR is 30% of 4,000 or $1,200. The difference between the TDSR (40%) and the GDSR (30%) represents the amount allowed for excess payments, i.e. car payments, credit card payments, furniture loan payments, etc.

When you contact the bank, do not say you are building your own home. The right time to tell them is when you apply for a mortgage and can demonstrate that you have the means and capability to complete the job.

To calculate the maximum mortgage you can qualify for, simply follow these steps:

(1) Multiply your monthly income by the bank's GDSR
(2) Deduct from that amount an allowance for taxes.
(3) Divide the amount left by the figure in the Monthly Payment Factor Table which corresponds to your interest rate and amortization period.
(4) Multiply this figure by 1,000 to find your maximum qualifying mortgage.

Example 1

Assume a GDSR of 30%, interest rate 8%, amortization over 25 years and estimated monthly taxes of $200.

The maximum mortgage an applicant would qualify for based on a gross income of $48,000, with monthly payments, would be calculated as follows:

Yearly income $48,000.
Monthly income 4,000.
Maximum P.I.T. 4,000. x .3 = 1,200.
Maximum P.I. 1,200. - 200 = 1,000.
Maximum mortgage is:
1,000 ÷ 7.63213 x 1,000 = $131, 030.

Example 2

Assume a GDSR of 30 %, interest rate 8.5%, amortization over 20 years and estimated monthly taxes of $150.00.

The maximum mortgage an applicant would qualify for based on a gross yearly income of $60,000, with monthly payments, would be as follows:

Yearly income $60,000.
Monthly income 5,000.
Maximum P.I.T. 5,000. x .3 = 1,500.
Maximum P.I. 1,500. - 150 = 1,350.
Maximum mortgage is:
1,350 ÷ 8.58559 x 1,000 = $157,240.

Calculating Your Monthly Principal and Interest Payment

To find the blended monthly payment of principal and interest on your mortgage, first find the blended payment factor in the table and then apply the factor as follows.

$$\frac{\text{Total Mortgage Amount}}{1{,}000} \times \text{Factor} = \text{Blended Mortgage Payment}$$

Example 1

Mortgage amount - $100,000
Interest Rate - 7%
Amortization - 20 years
Payment Factor - 7.69311 (for monthly payments)

Based on the given information the monthly principal and interest payment would be:

$$\frac{100{,}000}{1{,}000} \times 7.69311 = 769.31 \quad \text{(plus taxes for PIT)}$$

Example 2

Mortgage amount - $150,000
Interest Rate - 5%
Amortization - 25 years
Payment Factor - 5.81605

Based on above, the monthly principal and interest payment would be:

$\frac{150{,}000}{1{,}000}$ X 5.81605 = 872.41 (plus taxes for PIT)

TABLE # 1
BLENDED MONTHLY PAYMENT FACTORS FOR A LOAN OF $1.000.*

INTEREST RATES	10 YEARS	15 YEARS	20 YEARS	25 YEARS
5.0	10.58149	7.88124	6.57125	5.81605
5.25	10.70138	8.00909	6.70691	5.95918
5.5	10.82194	8.13798	6.84391	6.10391
5.75	10.94318	8.26790	6.98224	6.25022
6.0	11.06510	8.39883	7.12188	6.39807
6.25	11.18768	8.53076	7.26281	6.54742
6.5	11.31093	8.66369	7.40500	6.69824
6.75	11.43484	8.79761	7.54844	6.85050
7.0	11.54940	8.39249	7.69311	7.00416
7.25	11.68461	9.06834	7.83897	7.15919
7.5	11.81047	9.20514	7.98602	7.31555
7.75	11.93696	9.34287	8.13423	7.47321
8.0	12.06409	9.48153	8.28357	7.63213
8.25	12.19185	9.62110	8.43404	7.79229
8.5	12.32023	9.76158	8.58559	7.95364
8.75	12.44924	9.90294	8.73822	8.11614
9.0	12.57886	10.04519	8.89189	8.27977
9.25	12.70908	10.18830	9.04660	8.44450
9.5	12.83991	10.33226	9.20231	8.61028
9.75	12.97134	10.47707	9.35899	8.77708
10.0	13.10377	10.62270	9.51664	8.94487

* Based on interest being compounded semi-annually, not in advance.

To find the blended monthly payment of principal and interest on your mortgage, first find the blended monthly payment factor in the table and then apply the factor as follows:

$\frac{\text{TOTAL MORTGAGE AMOUNT}}{1{,}000}$ x FACTOR = MORTGAGE PAYMENT

WORKSHEET # 3
MORTGAGE CALCULATION

Applicant's gross yearly income	$_____________
Co-applicant's gross yearly income	$_____________
Other income	$_____________
	$_____________
Total Income	$_____________
Monthly Income	$_____________

Gross Debt Service GDS (30% of monthly income)		$_____________
Less estimated monthly property taxes		$_____________
Amount allowed for monthly principal and interest payments (P.I.)		$_____________
Divide by factor in Table# 1 corresponding to your amortization period and correct interest rate.	÷	$_____________
	=	$_____________
Multiply by 1,000 to find your **Maximum Qualifying Mortgage Amount**		(x 1,000) $_____________
Total Debt Service TDS (40% of monthly income)		$_____________
The amount allowed for excess debts is the difference between the TDS and the GDS. (10% of monthly income)		$_____________

Calculations

How A Small Increase In Your Mortgage Payments Can Save You Thousands Of Dollars

You can save thousands of dollars by shortening your amortization period. Illustrated are two alternative amortization periods, assuming a $120,000 mortgage and an 8% interest rate over the entire amortization.

	25 Year Amortization	20 Year Amortization	Difference
Monthly Payments	$915.86	$994.03	$78.17
Months Paying	300	240	
Total Cost	$274,758.00	$238,567.20	$36,190.80
Less Mortgage	120,000.00	120,000.00	
Financing Cost	$154,758.00	$118,567.20	**$36,190.80**

The additional monthly payments of $78.17 per month for 20 years represents an additional expense of $18,760.80. The savings in financing costs totals $36,190.80. You can choose an amortization period that coincides with your paying ability. Don't be coaxed into a 25-year amortization when you can afford a 17 year, 19 year or 22 year amortization. In the long run, your decision could mean a difference in thousands of dollars. The more you pay on your monthly payments, the more you will save on the total costs!

Interest rates will vary depending on the term you choose. Use worksheet #4 to work out various financing alternatives. Assume the same interest rate for the entire amortization period and note the differences in total costs. This worksheet will help you choose an amortization period and the term of your mortgage.

WORKSHEET # 4
CHOOSING A TERM AND AMORTIZATION PERIOD

MORTGAGE AMOUNT $ ______________________

TERM	RATE		AMORTIZATION 10 Years 120 Months	15 Years 180 Months	20 Years 240 Months	25 Years 300 Months
6 MONTH		PMT. Factor x Mortgage ÷ 1,000 = Monthly P. I. x No. of Months = Total Costs				
1 YEAR		PMT. Factor x Mortgage ÷ 1,000 = Monthly P. I. x No. of Months = Total Costs				
2 YEAR		PMT. Factor x Mortgage ÷ 1,000 = Monthly P. I. x No. of Months = Total Costs				
3 YEAR		PMT. Factor x Mortgage ÷ 1,000 = Monthly P. I. x No. of Months = Total Costs				
4 YEAR		PMT. Factor x Mortgage ÷ 1,000 = Monthly P. I. x No. of Months = Total Costs				
5 YEAR		PMT. Factor x Mortgage ÷ 1,000 = Monthly P. I. x No. of Months = Total Costs				
OTHER						

Estimating The Cost Of Your Home Prior To Construction

A rough estimate of the construction costs can be made by multiplying your proposed square footage by the estimated cost per square foot. Determining what to use for cost per square foot is the tricky part because it will vary depending on many factors; size of home, shape, style, quality and the amount of sweat equity or work you do yourself. For example: an 1,800 square foot two story house may cost $65.00 per square foot while a 1,800 square foot bungalow may cost $85.00 per square foot. The bungalow has twice the expense in the foundation footings, walls, floor and twice the roof trusses, roof insulation and shingles or other roofing materials.
To obtain a fairly accurate estimate, check with builders, realtors, your draftsman and any other sources involved in the building industry. Your cost per square foot will likely be lower because you are acting as your own general contractor, have no overhead costs and will probably do some work yourself.

Believe it or not, one of the homes we built, we had to build substantially larger to qualify for the financing. The larger home had obviously more square footage and with a lower construction cost per square foot, it made a greater difference between costs and appraised value. This greater difference allowed us to be financed under 75% which meant qualifying for the mortgage.
Once you have some idea about the approximate cost per square foot, you can estimate your maximum affordable house size. Remember, this method of estimating is to be used only as a ballpark figure for the cost of your home. It will provide you with a maximum house size, which you can use for a budget when shopping for plans. The only efficient way you can price out your home accurately is to get the blueprints and seek estimates for all the major work. This must be done prior to applying for a mortgage.
I remember spending $1,200.00 on house plans, pricing out a split level home only to discover that it cost $30,000 more than what I could finance. We scrapped the plan and started over. Hopefully now you will not waste $1,200 like I did!
On another occasion we added four feet onto the back of our home which increased the size by 130 square feet. The increase in construction costs was under $2,000 because few changes were made other than adding space to the family room and master bedroom. For example; no change to the kitchen, plumbing, heating, electrical, excavation, services and many more items.

The additional square footage was appraised (along with the total house) for $55.00 per square foot, adding value of over $7,000 to our home. All you need to do to see this is look at what items will change that are listed under the Cost Summary and Loan Calculation Form.

Complete **Worksheet #5,** the rough estimate sheet for the type of home you intend to build using estimated constructioin costs per square foot. What is the maximum house size you can afford to build? What style of home are you going to build? What is the maximum amount you should spend on a lot?

Note: The two storey home will likely have the lowest cost per square foot followed by the bungalow, then the bi-level. The split level (counting the top two levels only) will be the most expensive design.

WORKSHEET # 5
ROUGH ESTIMATE FOR THE COST OF CONSTRUCTION

Total Mortgage $__________

Add
Downpayment $__________

Extra Cash $__________

Subtotal $__________ $__________

Total amount of money available for construction $__________

Subtract
Total price of land $__________

Legal, insurance and interest fees
(approximately 2 to 4% of total house and land) $__________

Approximate cost of extra items
(examples; fireplace, jacuzzi, floor heating, decor extra lighting, any other elaborate or unusual items)
1/__________
2/__________
3/__________
4/__________

Subtotal (subtract from total available for const.) $ __________ $__________

Money available for building base house only **Total** $__________

CALCULATE WHAT YOU CAN AFFORD

HOUSE TYPE	TOTAL MONEY AVAILABLE FOR BASIC HOUSE (CALCULATED ABOVE)	DIVIDED BY ESTIMATED COST PER SQUARE FOOT BASED ON TYPE	EQUALS THE MAXIMUM HOUSE SIZE BASED ON TYPE AND BUDGET
BUNGALOW			
BI-LEVEL			
TWO STORY			
STORY & HALF			
SPLIT LEVEL			

Your Personal Budget Analysis

It is obvious that whether building a new home or a major renovation, there will be changes to your monthly obligations and the budget.

The first step is to list, as nearly as you can, where you are now with all your present expenditures. Complete a monthly budget analysis. Check the facts as they are today (analyse your present cost of living), take account of factors you cannot control (the fixed living expenses), eliminate the factors you can in some way control (other expenses), and plot a course. See Worksheet # 6. You may need a method of keeping records to know exactly where your money is being spent. Only 2% of North Americans follow a family budget and know how much it costs them to live each month!

The second step is to adjust those expenditures you can control to meet your goal (saving for a home). The main areas to save money are in living expenses and eliminating credit spending and subsequent monthly installments. Most financial problems an owner-builder encounters will be the result of poor budgeting.

The third step is to look into the future to see what your cost of living will be after you move into your new home. You may have to adjust your lifestyle to suit your new budget.

Now you are ready to plot a course, to establish priorities, and to ensure you will have sufficient cash flow while living in your new home.

In order to qualify for a mortgage, you may have to reduce your consumer debt payments from their current level to an amount allowable by the mortgage lender. (e.g., maximum 10% of gross monthly income). Check with your lender prior to applying for a mortgage.

WORKSHEET # 6
MONTHLY BUDGET ANALYSIS

INCOME		TOTALS
Net monthly income - Husband	$________	
Net monthly income - Wife	$________	
Family allowance	$________	
Interest income, investment income	$________	
Other income	$________	
Total Net Monthly Income	$________	$________
EXPENSES		
SHELTER		
1st. Mortgage (incl. taxes) or rent	$________	
Maintenance and repairs to property	$________	
Heating fuel	$________	
Power	$________	
Telephone (local and long distance)	$________	
Cable T.V. ----------------------	$________	$________
TRANSPORTATION		
Car operating costs, maintenance	$________	
Gasoline	$________	
Public transportation, parking --------------	$________	$________
LIVING EXPENSES		
Groceries, pet food	$________	
Clothing	$________	
Household incidentals; e.g. newspaper, magazines, etc.	$________	
Monthly recreation, school expenses, lessons, etc.	$________	
Personal allowance, lunches, coffee, haircuts, cosmetics	$________	
Entertainment	$________	
Alcohol, cigarettes, medication ------------	$________	$________
MONTHLY INSTALLMENT OBLIGATIONS		
Car loan	$________	
Furniture, appliances, other loans, etc.	$________	
Visa, Mastercard, American Express, other credit cards	$________	
Other	$________	
Other	$________	$________
SUBTOTAL OF EXPENSES ON THIS PAGE ----------		$________

WORKSHEET # 6 - Continued

IRREGULAR ANNUAL EXPENSES (÷ 12)		
Annual holiday	$________	
Furniture	$________	
Christmas, birthdays, etc.	$________	
Car insurance, licences	$________	
Fire insurance	$________	
Life insurance	$________	
Medical and dental	$________	
Membership fees, dues	$________	
Retirement savings plan	$________	
Other	$________	
Subtotal of all expenses (include page one subtotal) — —	$________	$________
Monthly income minus expenses		$________
Less amount for monthly savings plan		$________
Balance ————————————————————		$________

ADJUST COSTS OF SHELTER TO FUTURE COSTS OF LIVING		
Mortgage payments P.I.T.	$________	
Heating fuel	$________	
Power	$________	
Telephone	$________	
Cable T.V.	$________	
Water and Sewer	$________	
Home Maintenance	$________	
TOTAL ————————————————————	$________	$________

BUDGET FOR COST TO MOVE IN (Out of pocket expenses not planned in the mortgage)		
Moving expenses	$________	
Appliances	$________	
Home Maintenance, tools and equipment (lawnmower, weed trimmer, rake, shovel) ——————————	$________	$________

BUDGET FOR FUTURE ITEMS		
Landscaping	$________	
Curtains, blinds	$________	
Concrete driveway ?, deck, basement developement	$________	
TOTAL ————————————————————	**$**________	**$**________

6

SELECTING A SITE AND PURCHASING LAND

The amount of money you save by being your own general contractor will be directly related to the future appraised value of the home you build. It's not how much work you do, but how you plan the work that counts. As a general rule, placing your home in harmony with the land increases its desirability and the potential value of the property. Site and house design planning is crucial for saving money. Before placing a deposit on a lot, consider the fundamental principles of real estate appraisal discussed below and check the site selection checklist that follows. This information will help you understand the procedures of valuation. Then you will be ready to purchase your land.

Site Selection

1. Supply and Demand

The principle of supply and demand declares that the greater the supply of an item, the lower its price. When selecting your lot, consider the number of other vacant lots in the area. If the lots are not selling, they may be overpriced. The price should be at a point where supply and demand are equal. Use this principle to assess the future demand for your proposed home. An exclusive area with a limited number of homes will likely maintain a higher appraised value.

2. Principle of Substitution

How easily will your future home be substituted for another house by a potential purchaser? Suppose you plan to build a bi-level in an area where there is already an abundance of bi-levels. The future value of your home will be directly related to the market value of the other similar homes. You may not be maximizing your value for the area.

3. Principle of Highest and Best Use

Some sites are best suited for a particular style of home, depending on grades, adjacent buildings, and the nature of the area. The highest and best use for the site will be the one that compliments and produces the greatest value for the home you intend to build.

4. Principle of Anticipation

Land values will rise in anticipation of future benefits. The anticipation of a major oil facility to be built, the anticipation of winning the bid for the Olympics or the anticipation of high resale value from a newly developed subdivision are some examples of this principle. When selecting your site, anticipate future benefits (rising prices or satisfaction) and speculate on future market activity.

5. Principle of Balance

A decrease in value will result if the neighbourhood is not in good balance. Too many multi-family dwellings or too few services (stores, busses, schools, parks) will tend to reduce value. Before buying a lot in an urban area, check the regional planning office for the zoning of future developments and services.

6. Principle of Conformity

The home you intend to build must conform to the quality of other homes in the area in order to maintain maximum value. You would not build a log home in a subdivision where all the homes are aluminium and vinyl siding. Similarly, you would not build a large 2000 square foot estate home in an area where the average house size is 1000 square feet. The smaller homes will drag down the value of the larger one. A well-planned area will have a variety of styles all conforming to the existing standards of the area.

7. Attributes of the Lot

The owner / builder should be aware of the pros and cons attributed to various lots. Once you purchase your lot there will likely be no chance to change your mind and get a refund.

Corner Lots

ADVANTAGES

- Frequently larger
- Can place front of house in one of two directions
- Can obtain a wider front elevation by placing the house on the long side
- Possible to have a side drive
- Good potential for planning a future garage

DISADVANTAGES

- Usually more sidewalks to shovel
- Less privacy than interior lots
- More traffic and possibly more noise
- Possibly higher taxes
- Often lower appraised value
- May be harder to sell your home because corner lots are generally less desireable

Unless your proposed house design is best suited for a corner lot you are probably better off with an interior lot. An ideal location is a lake frontage or a view lot; however, they are often too rare and too expensive. Second choice is a lot on a quiet street close to a park or reserve area.

Sloped Lots

ADVANTAGES

- If the lot slopes down to the back, it may be suited to a front to back split design or a walk-out back basement entry
- May have better drainage depending on slopes of adjacent lots
- Gentle slope may add to appearance and landscaping

DISADVANTAGES

- May have to haul away excavation dirt, which will slightly increase costs
- If the lot slopes up to the back, it can cost extra for extension of basement walls and may limit your choises to split level designs
- If the lot slopes down to the back it, could cost extra for frost walls
- Tends toward a more detailed and expensive house design
- May limit recreational use of the yard
- More difficult to landscape

Narrow Lots

ADVANTAGES

- Possible cost savings
- Possible location advantages (closer to city center and less travel)

DISADVANTAGES

- Limited yard space and less privacy
- Limited to narrow house design plans which are often difficult to place front door and restricted in floor plan layout

The smart builder will do plenty of research before selecting a lot. By strategically planning a good location, you can increase future market value and savings. Many trade-offs will have to be made regarding the benefits of alternative sites and their costs.

Factors To Consider

The following factors should be assessed carefully before purchasing your land.

- Are the other homes in the area well-designed and constructed?
- Do the styles of homes in the area complement each other?
- Are satisfactory educational facilities available?
- Is there a nearby shopping centre, church, recreation area?
- Are the sewage, water, garbage, snow removal, lighting services satisfactory?
- Is there street sanding in winter and cleaning in the spring?
- Are public transportation facilities nearby?
- Is any development likely to take place that would depreciate the value of the property?
- Are there restrictions on the form of construction and architectural details (e.g. the size, type of dwelling, and colours)?
- Are the soil conditions good (no rock or loam) and is the lot level?
- Are there zoning restrictions and existing easements that could affect value?
- Is the neighbourhood improving in the process of growth and rising values or are homes depreciating?
- Is the property tax situation stable or are taxes likely to cause financial problems?
- Are the roads paved and services installed?
- Is the street short, quiet, and safe from traffic?
- Does the lot have a desirable shape to suit your plan?
- Will the finished grade be desirable for appearance and drainage?
- Are there flooding problems in the spring?
- Is there water, air or noise pollution?

There may be other points you can add to this list that pertain to your individual needs. For instance, will there be room for future development, and do the lots have lanes? Be sure to obtain answers to all these questions before you buy. (See Site Selection Checklist)

Purchasing Land

There are honest, reliable developers and happy, satisfied customers. Unfortunately, the reverse is also true. It's up to you to determine whether or not the property you want is a good buy and whether or not the seller keeps all promises. The only protection you have is through your own research of the developer's property report.

The first step is to analyse what your requirements are and exactly what you need. If this is done first, there should be little chance of being coaxed into something you don't want. You should have a good idea of the type and size of house you want to build. From this information you can judge the size of the lot and grading you require. List all your needs.

Your strategy should include keeping interest costs to a minimum. This is achieved by building on the lot as soon as possible or by winning concessions over the developer through negotiations. If the developer fails to budge on price, you might try to negotiate three months or more interest free on the outstanding balance.

Do your homework before you negotiate your land agreement. You will be better informed on all the details and have more information to negotiate with. Compare prices, do a title search, and get the facts on the lot on paper. Ask to look at the development plans and the property report if they exist. Obtain a copy of their standard offer to purchase and read it through.

Find out if you have to begin construction within a certain time period (e.g. subfloor stage within 365 days). It is usually to your advantage not to have a building commitment; however, a development with no building commitments may take years to complete.

Check the building requirements. In some controlled areas, the type of house, size, exterior materials and colours have to be approved by the developer prior to being issued a grade slip. The purpose is to maintain conformity and high value in an area.

When you finally make an offer, be specific on your demands! You have to know at this time how you intend to finance and eventually pay off the property. Some alternatives are listed in this chapter; however, you will have to check these alternatives with the land owner. **Any deposit you make on property should be placed in an interest bearing trust account for your benefit.** If you ask for this initially, you will avoid arguing afterwards about who gets the interest.

Negotiating

Negotiate price and everything else! Direct your energy at the most important issue first. If you can't agree on price, FORGET IT! There is no point discussing any of the terms or conditions of agreement. Only after price is within range do you begin, with the assistance of your lawyer, to work out all the other details. If the price doesn't work out the way you expected, then at least you haven't engaged the services of a lawyer and incurred legal fees.

Compare price to other lots in the area. Divide the total cost of the land by the total number of square feet to arrive at a cost per square foot figure. Now you have a measurement useful for making comparisons.

Take time to consider the agreement before signing. Don't sign any document unless you have given it at least 48 hours consideration.

When you have found a lot that meets all your requirements, bring the agreement to your lawyer before you sign. **Read your land agreement carefully then discuss any articles of concern with your lawyer.** He or she will advise you how to proceed with the seller's agreement. Once you sign it is a binding contract and you have to live with the terms regardless of how harsh they are. This is probably the most important transaction you will ever do. The cost of a good lawyer is well worth it.

Work out the entire deal in your mind first, then see your lawyer and discuss it before you sign. Your lawyer will provide insight into what's in the agreement and also what's missing from the agreement.

The lawyer's functions are as follows:

(a) Examine copy of agreement and provide comments to purchaser regarding terms and conditions.
(b) Do a title search, order tax certificate.
(c) Check to see that all utilities are paid.
(d) Do a general registry search on the original land owners. Check for judgments against them.
(e) Check for easements.
(f) Check to see if existing mortgages are assumable.
(g) Prepare a transfer of land.
(h) Provide mediation on behalf of the purchaser.
(i) Execute documents with purchaser and forward to developer's lawyer.
(j) Ensure all discharges from title are taken care of.
(k) Obtain and transfer funds from purchaser.
(l) Close outstanding documentation.

Avoid package deals. You want just the lot, not a package scheme where the land owner must also build your home.

Reduce any risk of the other party not following through on verbal promises. Ask which article in the agreement covers that particular promise.

Place a time limit on your offer. Deals can be lost if the other party has too much time to think it over.

Prepare how you intend to ask for something. The proper choice of words can make a difference in how it will be viewed by the other party.

Sometimes it pays to be a tough negotiator provided you know when to quit. Be prepared to make some concessions. You will likely underestimate what the other party wants.

Have provisions to legally cancel the agreement if things go wrong before the closing date.

Have your lawyer draw up a purchase agreement if the developer's standard is unfavourable to you. Your lawyer will have copies of agreements on file to work from.

After an agreement is reached, place documents in a safety deposit box for future reference.

Financing Alternatives

1. Developer Provides Financing

The preferable situation, when the vendor is going to provide financing, is a transfer of title with a small vendor take-back mortgage.

An alternative is an agreement for sale where the purchaser provides a smaller deposit (up to 25% of the purchase price) and the developer finances the balance for one year. Under these circumstances, the developer will hold title to the property until the land is paid in full. In the meantime, you have a registered interest against the property.

Example of terms:

- 10% down
- 15% on closing of agreement within 30 days
- Balance at 14%
- $1,000. security or damage deposit
- Pay out one year from closing

The advantages of this alternative are that you can tie up a lot with only a small amount of equity, and depending on the terms of the agreement, you may be able to start construction before the lot is paid for, allowing final pay out to come from first mortgage draw. This gives you better cash flow for construction.

Before signing any agreement, have the lawyer investigate the title and the position of the developer.

2. Complete Financing from Bank and Developer

In this case, the purchaser, who has no cash, uses his or her paying ability along with equity in existing properties to arrange a loan or second mortgage. The loan is used for a deposit on property. The remaining financing is provided by the developer until a mortgage is arranged or the purchaser sells an existing home.

With this method you can tie up a lot and begin planning for construction of your new home without having a dime in the bank. However, your interest costs increase, and you must pay out the lot within a specified time or lose your deposit.

3. Option to Purchase Agreement

An agreement between the purchaser and developer allows the seller to hold the land until a specified date and any other conditions are met (for example, proof of a mortgage).

With an option agreement, the purchaser is creating an interest in land and can register this interest in the form of a caveat against the property at the land titles office.

The lower the deposit, the better for the purchaser. In an option agreement, a purchaser who does not exercise the option will likely lose the deposit.

Some of the important considerations include:

- The option amount or down payment
- Balance owing
- Interest rate
- Pre-construction requirements
- Date of final payment

An option to purchase allows you to avoid losing a piece of property when you don't have the money for the entire property right away and gives you more time to plan, collect estimates, and apply for a mortgage. You also have a choice not to purchase just in case something goes wrong.

4. Making a Cash Offer

As obvious as it seems, a cash offer is another financing alternative which should be considered when possible. The purchaser will probably place a small initial deposit to be held in an interest-bearing trust account until all other terms and conditions are settled before the balance is paid.

The advantages of this option are the immediate transfer of title and a better position for negotiating. However, you may be using all your cash for the land leaving you to rely entirely on the bank to provide cash flow during construction.

5. Conditional Offer to Purchase

An agreement is frequently made subject to fulfillment of a condition (e.g. subject to receiving mortgage financing). The conditions and all the terms of the sale must seem palatable to both parties. They should be put in writing and taken to a lawyer to review before signing.

A conditional offer provides some flexibility to the purchaser - an out if you can't fulfill the conditions. However, the law is that a party cannot rely on the default of an attempt to fill a condition to get out of a deal.

Dishonest Sales Practices

1. Misrepresenting the Facts About Current and Future Resale Value

Before making a decision, a purchaser should thoroughly research the validity of the information provided.

The developer or sales agent may present general facts about an area's population growth, industrial and residential development, and real estate price levels as if they apply to your specific lot when in fact they apply to a distant city.

It may be difficult or impossible to recapture your initial investment in a future resale because the developer spent much of your purchase price on an intensive advertising campaign and commissions for sales agents. It is unlikely anyone would pay you more than you pay the developer, who probably holds extensive unsold acreages in the same subdivision.

2. Misrepresenting the Facts About the Subdivision

A developer or sales agent may offer false or incomplete information on the size of the property and the services supplied. The developer will probably not inform you of poor drainage, swamps, pollution, and claims or charges against title. Research the subdivision yourself.

3. Failure to Develop the Subdivision as Planned

The developer may make oral promises to develop the subdivision in a particular way. The promised attractions that influenced your purchase (park, playground, golf course, ski hill, tennis courts, swimming pool) may never materialize. They may be provided after a long delay, which in the meantime could affect your resale value or possibly your retirement plans.

4. Failure to Honour Damage Deposits

When you purchase a property, you may be required to pay an additional sum of money as security in case there is damage to sidewalks or services or you do not comply with development guidelines.

Under certain conditions this sum of money may never be repaid. In the agreement it could be held in an insurance fund and not repaid until every lot is developed, which could take several years.

5. Beware of the Contract

The contract you sign will be in favour of the developer. Read it through thoroughly to ensure all promises made by sales agents are in writing. If they are being truthful in their verbal promises, they should have no complaints about putting those promises in writing in the agreement.

6. Bait and Switch Tactics

Lots are frequently advertised at extremely low prices, but often those prices are just the starting prices of the unattractive smaller lots. Or the low price lots may all be sold and you are pressured to buy one that is more expensive. Be careful.

7. High Pressure Sales Tactics

Some sales agents may lead you to believe that the lots are selling rapidly and that you should make your choice right now. A developer may show you a large map of the area displaying sold stickers on most of the lots and give you a sales pitch about lot sales currently in progress. Don't be pushed to rush into a purchase you may later regret.

8. Know the Owner

A quick title search should be done to ensure the person you are dealing with actually owns the property and can convey clear title. Lots are frequently sold by sales agents who only hold an option to purchase, not clear title to the property. Ensure your security by researching details about the actual owners of the property.

9. Illegal Practices

Avoid any dealings with sales agents that could lead to fraud and a criminal offence.

Site Selection Guide

The purchasing or building of a new home is usually the largest single investment in a lifetime. The following check list will inform you of some of the more commonly overlooked items in selecting a building site. Keep in mind that every site and location creates an individual situation, therefore, consultation with local authorities and builders active in the area is recommended.

It may be difficult to obtain all of the advantages listed. However, the more existing improvements, the greater the loan and resale value of your home.

General Neighbourhood

Approach to Property
Does property access avoid blighted, unkept, sub-standard or industrial area? If not, in what relation to your site are these areas?
Surrounding Homes
Are nearby homes of equal or greater value than your proposed home?

Community Facilities

Distance to Grade and High Schools
Status of Schools
School schedules
Admittance availability
Controlled Crosswalks
Churches
Distance to and Extent of Shopping Facilities
Public Transportation
Fire Protection
Availability
Adequate fire protection is a requirement to obtain favourable fire insurance rates
Parks and Recreation Facilities

Utility and Improvements

Water
Check with district for actual availability for your particular site size of water service. Will the service be adequate for sprinkling systems?
Fire Hydrant
Does hydrant have adequate sized main? This factor will influence fire insurance rates
Electricity (overhead or underground service)
Gas
Street Lighting
Storm Sewers
Hard Surfaced Streets
Refuse Collection
Sanitation
If sewers are not available, will soil be approved for septic tank or cesspool? Check with sanitation authority in area.
Curbs and Gutters
Property Taxes
Sidewalks
Check and compare tax ratio of the property with rates in other available areas
Is the area free of tax burdens of annexed properties that do not offer benefits to your area?
(This situation may exist in rapidly developing areas)

Present Zoning Of Adjoining and Surrounding Property

Zoning for Lot Size
Minimum width and depth
Zoning for Use
Zone Changes
Check with local planning commission about recent zone change proposals in the area that may influence your property
Property Restrictions
Restrictions of building size, set backs, side lines and public easement for power, sewage, water, etc. Be aware of industrial or commercial influence and expansion in the neighbourhood of your site.

Site Disposition

Drainage
Is the lot well drained? Check with neighbours if in doubt

Sub Soil
Type: Rock, Clay, Sandy, other
Where septic tank or cesspools must be used, the nature of sub soil is important

Elevation of existing sewer laterals
Sufficient depth to assure gravity drain to sewer, particularly if any plumbing is located in the basement

Slopes
Steep slopes require earth removal and retaining walls. Reverse slope may necessitate expensive fill

Natural Features

Desirable view
Trees, shrubbery
Lack of natural advantages often require extensive Landscaping

Weather
Wind direction, sun, fog, rain, snow
Become informed of weather indigenous to the site

Orientation

Consider Sun on Houses
Sun is desirable
North light is best for even lighting, which is most desirable in active living areas.
Southern exposure is beneficial for winter heat, if roof overhang is adequate for summer heat control.
Patio area should be arranged to take advantage of both sun and shade
Take advantage of the sun, views and prevailing winds in the rooms where these would be most desirable

Location of House
Relate the location of your home to other existing homes in the neighbourhood. Avoid placing picture windows opposite picture windows across the street Leave all trees and shrubbery until stake-out of house, then remove if necessary. Set grade of house to properly drain yard area and rain drains. Allow ample space for outdoor activity requirements and for planting areas on south side of the property. In all situations, privacy and beauty can be gained by proper planting and landscaping.

7

MORTGAGE APPLICATION

Financing

Arranging financing may be the most difficult aspect of building your own home. It is a barrier for many, even those who are employed full-time in the construction industry.

The goal here is to help you through this barrier by preparing you for mortgage and interim financing interviews. Once you have all the required information and can answer all the related questions, your chances for loan approval will be greatly increased. The banker will look for ways to assist you rather than ways to turn the loan down.

Definitions

To begin you must have an understanding of the following terms, which will be referred to throughout this chapter.

1. Mortgage

A mortgage is simply a loan secured by real estate. The security is in the form of a contract detailing payments by the borrower (mortgagor) of a given sum of money plus interest to the lender (mortgagee).

2. NHA Insured Mortgage (Canada ONLY)

Through the National Housing Act, the federal government will guarantee a mortgage for the bank or other approved lender. It is used for existing homes and new residential construction. However, the financing must be arranged prior to building. Using NHA allows the lender to loan up to 90% of the appraised value, which is substantially higher than conventional mortgages. Builders can obtain progress advances based on Canada Mortgage and Housing Corporation inspections as the construction proceeds.

3. Conventional Mortgages

Mortgage loans, other than those involving NHA and loans for less than 75% of the appraised value, are commonly called conventional mortgage loans. These loans are available through banks, life insurance, trust, and other loan companies.

4. Open Versus Closed Mortgages

The closed mortgage does not allow prepayment faster than has been agreed upon. In an open mortgage you can pay off or pay down the mortgage prior to the end of its term. Closed mortgages prevent the borrower from paying back too quickly a mortgage that is earning high interest for the lender.

5. Draw Down Mortgage

In a draw down mortgage, the borrower receives portions of the total mortgage, called progress advances, at various stages of construction after a satisfactory inspection by the lender.

6. Completion Mortgage

A completion mortgage is taken out after the home is constructed and appraised. The benefit of this mortgage to the lender lies in the security of an already built home. However, the builder requires an alternative source of financing to construct the home.

7. Amortization

The amortization of a mortgage refers to the method of repaying the principal. The most common is the blended payment of principal and interest in equal monthly sums. The amortization period can vary depending on the amount borrowed and the payments one can afford. The longer the amortization period, the lower the payments and the more you can borrow. The most common period is 25 years when borrowing the maximum qualifying amount.

8. Term

The term is the period for which the mortgage is written, and it frequently refers to the length of time the interest remains fixed. It is usually one to five years depending on the lender. With fluctuating interest rates, shorter terms and even floating rates have become popular.

9. Mortgage Insurance Fee (MIF)

The MIF is a fee paid by the borrower as a portion of the mortgage (usually 1%). This fund is to pay the mortgage company if there are losses due to foreclosure. **The fund does not in any way insure the borrower.** Separate mortgage or life insurance can be taken out for the borrower's benefit.

10. Gross Debt Service Ratio (GDSR)

This ratio represents the maximum percentage of the applicant's monthly income which the lender will allow for monthly principal, interest, and tax (PIT) payments. For example, a yearly income of $48,000 or $4,000 per month, with a 30% GDSR would mean a maximum PIT of $4,000 x .3 or $1,200.

11. Total Debt Service Ratio

This ratio refers to the maximum an applicant is allowed for PIT plus all other debt obligations. For example, a yearly income of $48,000, or $4,000 per month, with a 40% TDSR would mean maximum total debt payments of 4,000 x .4 or $1,600. The difference between the TDS of $1,600 and the GDS of $1,200 represents the amount allowed for excess payments, i.e., $400 for car payments, charge card payments, etc.

12. Interest Adjustment Date

The interest adjustment date is when all accrued interest from mortgage progress advances on a draw down mortgage is calculated and deducted from the final advance. The interest adjustment date usually is the end of the month following a completed inspection of the home.

13. First Payment Date

The first payment date is when the first mortgage payment is made. On a draw down mortgage, this date is one month following the interest adjustment date. The owner-builder can be living in the new home up to two months before making the first mortgage payment.

14. Pre-authorized Chequing

A lender usually will require your authorization to automatically debit your chequing account each month for your mortgage payments.

15. Pay out Penalty

The pay out penalty is a clause in a closed mortgage requiring the borrower to pay a penalty (usually three months' interest) if the mortgage is paid out or renegotiated before the end of its term.

16. Survey Certificate

A survey certificate is provided by the surveyor after an inspection of the foundation walls. The certificate verifies that the house was located in accordance with the zoning requirements of the property. It is required by the lender to ensure their security prior to advancing funds and usually by the town or city office to ensure the builder conforms to their by-laws.

17. Mechanic's (or builder's) Lien

A lien is a sum of money (usually 15% of the construction cost) that is held back until 35 days after final inspection of the property. The purpose is to ensure the security of the mortgage against any liens placed by subtrades who were not paid.

18. Interim Financing

Interim financing is a construction loan that the builder may require to cover expenses until the mortgage advance is received or until the home is complete and a mortgage is found.

19. Interest

Interest may accrue in four ways while the house is under construction: on mortgage draws, on interim financing advances, on land payments to a developer, and on overdue credit accounts with suppliers and trades.

20. Vendor Take-Back Mortgage

If you have a house to sell, you can increase the potential number of buyers by offering a vendor take-back. This means that the buyer gets a mortgage from you rather than from a bank. If the interest rate you offer is lower than conventional interest rates, you can probably get a better price for the house.

There are several precautions to take when offering a vendor take-back:

(a) Make sure the buyer's down payment is significant (at least 20% of the purchase price of the house). That will discourage the buyer from walking away from the property later on.

(b) Make a credit check on the purchaser. The mortgage is only as good as the people who pay the mortgage.

(c) Be certain to use a lawyer to draw up the documents. You want to be sure that you have security in case you have to foreclose. The lawyer will see that the mortgage is registered properly with the provincial land registry office.

(d) The take-back mortgage should be a short-term one: perhaps one to three years. At the end of the term you can set a new rate of interest if desired.

21. Mortgage Brokers

Mortgage brokers may be able to help you overcome problems you might encounter in dealing with bank managers and loan officers. Brokers are professionals who specialize in the negotiating, obtaining, securing, and closing of mortgage transactions.

Brokers usually represent a number of mortgage institutions, such as trust companies, finance companies, insurance companies, and banks. For this reason they are regarded as agents. Depending on their resources, brokers may be able to arrange financing faster by acting the same day it is presented. They can approach several lenders in the time it would take you to arrange an appointment with one potential lender.

Mortgage brokers are required by the Mortgage Brokers Regulation Act to disclose all terms, fees and commissions to the borrower. Included are legal fees, appraisal fees, mortgage insurance fees, and commitment fee to the agent or broker, which is usually a percentage of the mortgage. These fees are deducted from the principal amount or the mortgage is increased to cover them.

Since it may cost you more to get a mortgage through a broker, you should investigate several banks and other lending institutions first.

Always Shop Around For The Best Deal

Find out what types of mortgages are available, who has the best interest rates, the options that best meet your needs, and what prepayment privileges are available at what cost to you. To begin, here are some sources to consider.

(a) National Housing Act loans from chartered banks
(b) National Housing Act loans from life insurance companies, loan corporations, and trust companies, which are listed as "approved lenders" by CMHC
(c) Conventional loans from banks, life insurance companies and trust companies

(d) Private estates administered by trust companies and lawyers
(e) Private individuals and finance companies

Presenting Your Application

Your key to success is in your preparation and in making a good presentation. Lenders are asked for many more loans that they grant; they must be given the fullest possible information about the property and the borrowers. If you arrange an interview without sufficient preparation, the lender will lose confidence in you as a builder.

The only time you should approach a lender with your application is when you are ready to talk ***dollars***. Don't be a "walking generality."

What Criteria Must You Meet?

At the least, you must meet the following criteria:

(a) **Ability:** your ability to make mortgage payments, determined by your GDS and TDS ratios.
(b) **Stability:** your residence stability as well as your job stability determined by your employment verification and the stability of that industry.
(c) **Security:** based on the bank's appraisal of your house plans compared to the amount of money you require.
(d) **Age:** your age now as compared to the length of amortization you are applying for.
(e) **Health:** your physical condition as it might affect your ability to repay.
(f) **You:** first impressions, your attitude, your preparation and planning (the most important of all criteria and the biggest reason for most declined loans).

Helpful Tips For Your Loan Interview

(a) Always make an appointment first.
(b) Introduce yourself to the lending officer.
(c) Both husband and wife should attend when a couple is applying.
(d) Leave the children with a sitter. The lender wants all your attention when you are making important decisions involving thousands of dollars. Be an active listener to everything the lending officer says.
(e) Dress accordingly, first impressions are important. Approach the lender as though you are a business manager.
(f) Be prepared; bring all the required information (see next section).
(g) Always be polite and courteous even if you receive negative feedback.
(h) Be positive and optimistic at all times. Show motivation and determination to do what you set out to do.
(i) If this is your first home and you have little experience, don't remind your lender of this when you are attempting to gain his or her confidence.

How A Draw Mortgage Works

For the owner-builder, a draw mortgage is the most common method of receiving mortgage money. Unless you have a down payment of at least 25% of the appraised value, you will likely go to a bank, which will process your loan through the Canada Mortgage and Housing Corporation. CMHC will control the inspections and authorize the advance of funds. This process ensures the security of the lender by not advancing funds until work has already been completed. Provincial mortgage corporations may have slightly different procedures. As you will see, guidelines, qualifications, and documentation required may vary depending on the lender's policy.

The basic procedure on a draw mortgage is as follows:

(a) Begin construction using your own money or an interim financing loan from the bank.
(b) Call the lender or CMHC for an inspection once the subfloor stage is reached.
(c) When inspection is approved and request for first advance of mortgage funds is made, approximately 16% is advanced.
(d) Lawyer (acting for lender) receives first advance from lender and usually the land advance as well at this time.
(e) Lawyer searches the title for any liens, pays out the balance owing on the land, registers title securing lender's interest, processes mortgage documents with owner-builder, and advances the first progress advance to the bank.
(f) At this stage you are required to bring to the lawyer the survey certificate and proof of the required fire insurance (minimum amount to be specified by the lender).
(g) Bank receives money from lawyer, deposits it in your account to apply directly against outstanding interim financing loan.
(h) Steps are repeated at different stages of construction.

The following diagram illustrates how progress advances are made.

FLOW OF FUNDS AND COMMUNICATION DURING CONSTRUCTION

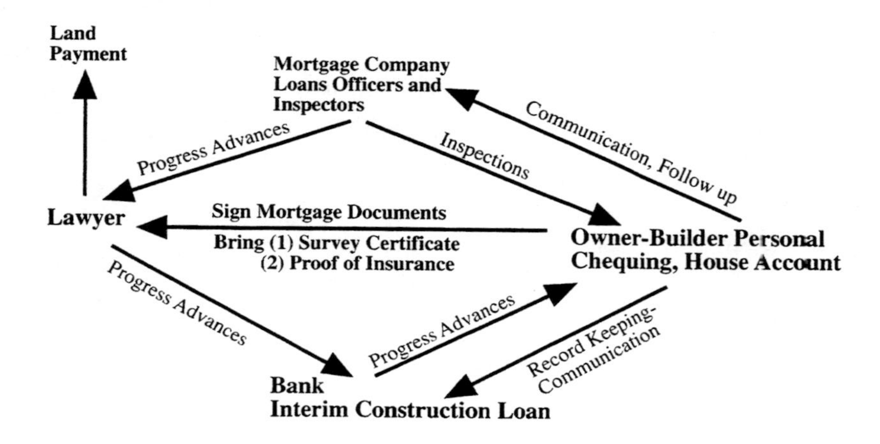

Following is a breakdown of how and at what stages mortgage proceeds are advanced, based on a new construction mortgage administered through Canada Mortgage and Housing Corporation.

Draw Mortgage Outline Of Payments

Line (1) TOTAL MORTGAGE AMOUNT $ ________
Subtract
Line (2) -Mortgage Insurance Fee (if req'd) $ ________
-Land value less land deposit (A seperate land advance is made to pay off any
Line (3) remaining balance) $ ________
Line (4) Subtotal $ ________ $ ________
Line (5) TOTAL PROGRESS ADVANCES $ ________

INSPECTIONS AND ADVANCES

1. **Foundation and Subfloor,** prior to backfill (approximately 16% complete)
First advance 16% of above progress
Line (6) advances (line5) $ ________
Line (7) Subtract 15% Mechanic's Lien Holdback $ ________
Line (8) Equals Total First Advance $ ________ $ ________

2. **Structural and Insulation**
After poly and before drywall
(approx. 40-45% complete)
Second Advance 40-45% of above
Line (9) progress advances (line5) $ ________
Line (10) Subtract First Advance (line 6) $ ________
Line (11) Subtotal $ ________
Line (12) Subtract 15% Mechanic's Lien Holdback $ ________
Line (13) Equals Total Second Advance $ ________ $

3. **Intermediate**, base coat of paint, ready for cupboards,
(approximately 70% complete)
Third Advance -70% of the above
Line (14) progress advances (line 5) $ ________
Line (15) Subtract First and Second Advances (line 9) $ ________
Line (16) Subtotal $ ________
Line (17) Subtract 15% Mechanic's Lien Holdback $ ________
Line (18) Equals Total Third Advance $ ________ $ ________

4. **Occupancy Inspection**, 100% complete
(Possible to have deficiencies which would require an additional final inspection)
Fourth Advance -100% of above
Line (19) Progress advances (line 5) $ ________
Line (20) Subtract First, Second and Third advances (line 14) $ ________
Line (21) Subtotal $ ________
Line (22) Subtract deficiencies holdback (bank may hold 200% of estimated value) $ ________
Line (23) Subtotal $ ________
Line (24) Subtract 15% Mech. Lien Holdback $ ________
Line (25) Equals Total Fourth Advance $ ________ $ ________

5. Final Inspection (if required)
Deficiencies are corrected

Line (26) Advance amount held for deficiencies $ ________

Line (27) Subtract accrued interest on all advances projected to the end of month or the interest adjustment date $ ________

Line (28) Equals Final Advance $ ________ $ ________

6. Release Mechanic's Lien Holdback
45 days after final advance
(check with Lawyer, the actual time will vary among Provinces/States)

Line (29) 15% of all previous advances $ ________

Note: Some legal fees will be deducted from advances. The amount and timing of this payment will vary from lawyer to lawyer.

Draw mortgages for new construction can also be otbained through regular lending institutions which will lend up to 75% of the appraised value. The lenders hire an independent appraiser to inspect the property during various stages of construction. Based on the appraisal report, the lender advances funds knowing there is security in the home.

Policies will vary with every lender and every appraisal company. When you apply for a mortgage, find out when and in what proportions you will receive mortgage progress advances. This information will benefit you in determining the amount of interim financing you require. An example of an inspection report is provided here for your information. The percentages shown can be a valuable benchmark for you to compare to when estimating for a new home.

(Example is based on a two story home with construction costs of $163,000)

EXAMPLE OF APPRAISER'S INSPECTION REPORT

(PROGRESS REPORT FOR MORTGAGE ADVANCES)

DATE: ____________ FILE NO. ____________
CLIENT: ____________ INSPECTION NO. ____________
ADDRESS: ____________ INSPECTION FEE. ____________
LEGAL: ____________
NAME OF BUILDER: ____________

		% Completed 1st	2nd	3rd	Final
Plans and permits	1.96%				
Survey	.30%				
Excavation, backfill, trenching, etc.	2.20%				
Basement concrete	3.37%				
Weeping tile, dampproofing	.67%				
Cribbing (basement forms)	1.96%				
Basement floor, concrete, and gravel	1.84%				
Lumber, framing materials & trusses	17.79%				
Stairs	.36%				
Framing Labor	5.21%				
Doors and windows and garage door	6.13%				
Roofing	2.70%				
Stucco, siding, bricks	5.52%				
Fascia, soffit, eaves	1.53%				
Plumbing, marble fixtures, tub, etc.	5.82%				
Heating	3.07%				
Electrical, intercom	4.74%				
Insulation and drywall,material & labor	7.11%				
Kitchen cabinets	4.36%				
Floor coverings	5.83%				
Tiles, bathroom, foyer, Mat. & Labor	1.22%				
Electrical fixtures	0.74%				
Painting and staining material and labor	3.68%				
Interior finishing	4.79%				
Interior finishing carpenter labor	2.45%				
Front steps	0.37%				
Driveway and walk	2.45%				
Extras, fireplace, appliances, etc.	2.76%				
Total	100.00%				

General contractor's overhead & profit add 12%

Amount of mortgage $ ________ less holdback $ ________ = $ ________
Less previous advances $ ________
Unadvanced balance $ ________
Less cost to complete $ ________
Advance now authorized $ ________

INSPECTOR'S COMMENTS

____

____

SIGNATURE ____________________________

Your Mortgage Documents

Follow the list of required documents in Worksheet #7. Most of the documents can be prepared prior to your interview, which will reduce approval time. To simplify matters, the checklist is broken down into three categories of information:

(a) Personal and financial
(b) Costing and loan calculation
(c) House plan information

1. Personal and Financial Information

Application Fee: Bring your checkbook with you and be prepared to pay for an appraisal fee prior to receiving your mortgage approval. Try to have the application fee waived. (See money saving ideas - no fees)

Application Forms: All lenders have their own application forms which they will fill out during your interview and which will require your signature. To make things easier for your mortgage officer you can prepare a personal statement using Worksheet #8. This saves the lender time and also reduces the number of probing personal questions you will be asked.

Income Verification: Have your employer supply you with a confirmation of employment on the company's letterhead or use the form in Worksheet #9. The letter should confirm your length of employment, present income, job position, and, if an increase in income is expected in the near future, how much and when.

If you have changed jobs recently, obtain a letter from your former employer, as a two-year job history may be required. Be as specific as possible; your income will determine your paying ability and the amount of mortgage you can receive. If you are self-employed, you will be required to provide your business income statement and balance sheet for the past one or two years of operation.

WORKSHEET # 7

CHECKLIST FOR MORTGAGE APPLICATION

Required Documents CHECK	Extra Items (May be Required) LIST	
		Personal and Financial Information
❒		Application fee
❒		Application form (for final interview)
❒		Personal statement (see worksheet #8)
❒		Income verification (see sample form, worksheet #9)
	❒	Income tax return (if bank requests it)
❒		Financial statement and statement of equity (worksheet #10)
❒		Photocopies of cheques showing deposits made on land, other
❒		Statement from bank verifying amount for deposit
❒		Copy of land agreement with legal and civic address
❒		Sample cheque for pre-authorized mortgage payments
	❒	Discharge from bankruptcy or explanation of any poor credit
	❒	Divorce papers
	❒	Copy of marriage certificate
	❒	Copy of social insurance (security) cards
	❒	Directions outlining how to find the property
		Costing and Loan Calculation
❒		Cost Summary and Loan Calculation form (see worksheet #11)
❒		Supporting written estimates from trades and suppliers
	❒	Contract or consulting agreement with builder (if hired)
		House Plan Information
❒		Two complete sets of house plans (give only one initially)
❒		Specifications of materials to be used (see spec. checklist)
❒		Plot plan or site plan (attach a copy to each set of plans)
❒		Grade slip (attach one copy to each set of plans)
	❒	Engineered approved plans for a preserved wood foundation
	❒	Engineered approved truss statement (from supplier)
	❒	Engineered approved beam details or other design details
	❒	Heat loss statement and layout (from heating contractor)
	❒	Well or septic system plans

WORKSHEET # 8

PERSONAL STATEMENT

APPLICANT		CO-APPLICANT
____________________	Full Name(s)	____________________
____________________	Age	____________________
____________________	Drivers Licence No.	____________________
____________________	Social Insurance (Security No.)	____________________

Marital status__________ No. of dependents including spouse______

Home address____________________________ How long__________

Previous address____________________________ How long__________

APPLICANT		CO-APPLICANT
____________________	Employer's name	____________________
____________________	Address	____________________
____________________	Type of business	____________________
____________________	Position	____________________
____________________	Telephone	____________________
____________________	Years there	____________________
____________________	Previous employer	____________________
____________________	Years there	____________________

Name of bank	Address	Telephone	Account No.
____________	____________	__________	__________
____________	____________	__________	__________
____________	____________	__________	__________

Name of landlord____________________________ Telephone ____________

Nearest relative____________________________ Relationship____________

Address____________________________ Telephone____________

Credit cards	Account no.
______________________________	____________
______________________________	____________
______________________________	____________

Other references____

____

I hereby certify that the above information is, to the best of my knowledge, complete and accurate.

Signature______________________________

WORKSHEET # 9

INCOME AND EMPLOYMENT VERIFICATION

Applicant

____

To: (Employer and Address)

____

I have made application to a mortgage company and I am required to provide verification of my annual income at present rate of remuneration. Would you please provide this information by completing and signing the lower portion of this form. It is important that the information provided be as accurate as possible.

Your early attention would be appreciated.

Date________________ Signature of employee________________________

EMPLOYER'S VERIFICATION

The following information is provided in strict confidence as requested by the above employee.

Period of employment ____

Present position ____

Annual salary/wages ____

Additional annual earnings for:

Overtime $____________ Bonus ____________

Commissions $____________________________

Prospects of continued employment rate: Excellent____ Good____ Fair ____

Comments ____

Date ______________________________ Signature of Employer ____________________

Position ________________________

Financial Statement and Statement of Equity: Preparation of a financial statement (Worksheet #10) in advance will eliminate fumbling and possible embarrassment during a mortgage interview. Both husband and wife will be on the same track, so you won't give conflicting information, which could easily result from not planning and communicating. There should be no doubt in the lender's mind about the accuracy of your financial statement. Tell the truth! If the lender finds conflicting information through a credit check or other means, your application will be denied almost automatically. The lender will not use this form, but will appreciate your preparation and presentation of all the required information. **You should not complete a bank application form during your submission because you already have all the information they will require.**

Verification of Down Payment: Before lenders will consider taking any risk, they want to see you invest and take some of the risk first. You should prepare verification of all your cash input prior to applying. This may take several forms:

(a) Copies of cheques showing deposits made on land and items purchased, e.g. blueprints, building permits, etc.

(b) Letter from your bank verifying amounts on deposit (if other than the bank you are applying to)

(c) Verification of equity in present home which you will be investing in the new construction (a confirmed sale agreement and evidence of a sufficient deposit to guarantee the buyers won't walk away)

(d) Letter from anyone else detailing any assistance they will be providing

The verification of down payment should be attached below your financial statement and statement of equity because they verify those items marked as liquid assets.

Don't be discouraged if this seems like a lot of work. It is easy once you know where to go and what information to collect. Remember, planning and management are very important to your building success.

Copy of Land Agreement: Security is another criterion for loan approval. Lenders ensure their security by lending less than the appraised value and by registering the mortgage with the local land titles office. To be best prepared you should provide your lender with a copy of the land agreement and a certificate of title showing all charges, liens, and interests on the property. With this information the lender can easily see who has to be paid off in order for them to register first title to the property.

2. Costing and Loan Calculation

Cost Summary and Loan Calculation Sheet: The cost summary (see Worksheet #11) is your proof that you have completed all the necessary costing and research to justify your budget for the home you intend to build. Check with your lender on their mortgage insurance fee and ensure you have a sufficient deposit. Next, complete this form to the very bottom. You are now ready to talk dollars.
The cost summary sheet is a summary of a more detailed listing you will use for estimating purposes (see Chapter on Estimating).

WORKSHEET # 10

FINANCIAL STATEMENT AND STATEMENT OF EQUITY

Name of Applicant(s) (in full)____

ASSETS AND LIABILITIES

ASSETS:		LIABILITIES:	
Cash	$________	Banks	$________
Deposit paid on property	$________	Finance Company	$________
*Stocks and bonds	$________	Credit cards	$________
Equity in present home (closing date of sale) ____________19_____	$________	Stores	$________
		Personal loans	$________
*Real estate	$________	other loans	$________
*Automobiles	$________		$________
Personal effects, household goods, etc.	$________		$________
Retirement savings plans	$________		$________
*Other	$________		$________
TOTAL	$________		$________

Details of assets marked*____

____

____

DETAILS OF ALL LIABILITIES SHOWN ABOVE

To whom payable Name and address	Date of origin	Balance owing	Monthly payments
	TOTAL	$________	$________

Verification that equity required, as shown above, is available must accompany this statement (i.e., verification from a bank of balance on deposit; copies of receipts for downpayment and other items paid)

CERTIFICATION

I hereby certify that the above information is, to the best of my knowledge, complete and accurate.

Applicant(s) Signatures ________________________

____

Date ________________________

WORKSHEET #11

COST SUMMARY AND LOAN CALCULATION

Builders Name(s) ________________________________

Preparation

- Blueprints, site plans, survey, etc. (1) $ ________

Foundation

- Excavating, backfill, trenching, grading, loaming (2) $ ________
- Basement materials (forms, concrete footings, concrete/preserved wood walls, weeping tile, dampproofing) (3) $ ________
- Basement labour, cribbing concrete or framing walls (4) $ ________
- Basement floor (gravel, concrete and finishing labour) (5) $ ________

Structure

- Framing materials (floor joists, sheathing, wall studs, roof framing, trusses, etc.) (6) $ ________
- Framing labour (7) $ ________
- Doors and windows (8) $ ________
- Roofing material and labour (9) $ ________
- Siding, brick and/or stucco, battons, trims, etc. (10) $ ________
- Fascia, soffits and eavestroughing (11) $ ________

Mechanical (all items supply and install)

- Plumbing (12) $ ________
- Heating (13) $ ________
- Electrical (14) $ ________

Finishing (all items summary of material and labour)

- Insulation, caulking and poly vapour barrier (15) $ ________
- Drywall (material and boarding, taping, sanding) (16) $ ________
- Kitchen cabinets, bath and laundry room vanities and counter tops (17) $ ________
- Floor coverings (underlay, carpet, linoleum, tile) (18) $ ________
- Ceramic tiles, Bathroom accessories (19) $ ________
- Electrical fixtures (all lights, bulbs, luminous panels) (20) $ ________
- Painting (primer, paint, stain, lacquer) (21) $ ________
- Interior finishing (railings, casings, baseboards, hardware)(labour) (22) $ ________
- Exterior parging (23) $ ________
- Steps, sidewalks, garage pad and driveway (24) $ ________
- Extras (i.e. appliances, fireplace, garage door, etc.) (25) $ ________

Services and permits (26) $ ________
Builders costs (general contractors or consultant fees) (27) $ ________

CONSTRUCTION COSTS ________

Contingency Factor (3-5%) $ ________
Land (incl. interest, taxes, security deposit) $ ________

TOTAL (finished home and land) $ ________
Legal and interest costs (2-4% above total) (28) $ ________
SUB TOTAL $ ________
Mortgage insurance fee, appraisal fee $ ________

Other $ ________

TOTAL CONSTRUCTION COST $ ________

LESS: Downpayment (deposits paid, items purchased, cash, other equity) $ ________

TOTAL LOAN / MORTGAGE REQUIRED $ ________

Supporting Estimates From Trades and Suppliers: Signed written estimates for all the major work (items (1) to (18) on the cost summary sheet) should be ready to present to your lender just in case you are questioned on the validity of your estimates. In addition you can prove that you have qualified tradespeople doing all the major work. Express a manager's point of view rather than discussing all the work you intend to do yourself. Once your loan is approved your lender cannot hold you to your estimates (for they are only estimates) and perhaps you might find better estimates and also do a little more work yourself. At least you will have sufficient funds to contract out everything just in case it is required.

Contingency Factor - A Built-in Cash Reserve: It's far better to borrow a few thousand extra than to run short and have to explain to your banker why you need a second mortgage or personal loan. Adding in a contingency factor of 5% of the construction costs (show a figure minimum of $5,000.00) may help to cover you for cost increases and forgotten items. A smart builder will use this money only for unexpected occurrences to ensure there is enough to do the job.

Legal and Interest Costs: Legal fees and accruing interest are major costs in any construction project. The longer the construction period, the higher the interest costs. You will find that 2 to 3% of the cost of the house and land is usually sufficient to cover a three-month construction period assuming mortgage rates are below 10%. Deciding whether this is enough for your budget will depend on several factors. These include your cash input, interest on land prior to building, length of construction time, date of your first payment, use of your lines of credit, and the actual rates you will be paying.

The general contractor must be able to comprehend how these costs are actually calculated. Having this knowledge will enable him to make trade-off decisions about those things he can influence, e.g., scheduling construction, paying of accounts, and arranging lines of credit.

For the most part, your rate of interest and the mortgage amount required will remain fixed. Postponing your payments on accounts and arranging lines of credit will certainly help you save interest. But, accounts eventually have to be paid. This leaves SCHEDULING CONSTRUCTION, which above all will have a definite bearing on the accruing interest costs you will have to pay.

The mortgage draw interest costs can be estimated for your construction period. The example that follows illustrates how the interest is calculated based on a three-month construction period. **Always remember, if it takes you twice as long to build the house, the interest costs can double!** For this reason you should study the chapter on job scheduling very carefully.

If you purchase land from a developer or realtor months in advance of construction, you may have to budget for extra accruing interest on an outstanding balance. You should attempt to eliminate or reduce this interest as much as possible.

The interest on your interim financing will be impossible to estimate 100% accurately prior to construction. Your balance owing will fluctuate daily as you pay subtrades and receive mortgage advances. To arrive at a useable figure, take one-half of the amount of interim financing you require (see Interim Financing), multiply by the going rate for interim loans (Try for prime + 1), then divide by 3 to cover you for 1/3 of the year or 4 months. Use four months because your final advance will not be received until some time after you move in. This assumes you will build the home in three months.

Figure out your legal and interest costs on Worksheet #12 and your accruing interest on mortgage progress advances on Worksheet #13. Remember, financing is the key to your home, this material requires your time and patience to understand.

3. House Plan Information

Complete Sets of House Plans: Depending on your lender you will be required to submit either two or three complete sets of house plans. Check with your lender and be certain your plans include all the right specifications and necessary details (e.g. metric or imperial measurements). Note: Offer only 1 set of plans until you receive approval.

Plot Plan: A plot plan is a scale diagram showing the location of the house on the property conforming to local regulations established by the city. (See sample plot plan). Check with the local building inspection department or your land developer to obtain the minimum house setback and side yard requirements.

Grade Slip: In a controlled subdivision the elevation of your house and property grades will be preplanned and regulated by the land developer . Once your plan is approved, you will be issued a grade slip (See sample grade slip). Attach a copy of your grade slip to each set of plans used for mortgage and building permit applications. Your surveyor will also require a copy to set the required cuts on the stakes for the excavator.

The Credit Check

All the personal and financial information you leave with your lender will be verified with the credit bureau, employer, landlord, banker, and other sources. The credit bureau maintains a history of all information reported by various lenders and retail stores. If you had any bad credit reports, it is best to explain the situation during your interview or check first with the credit bureau to see if it has been removed.

SAMPLE PLOT PLAN

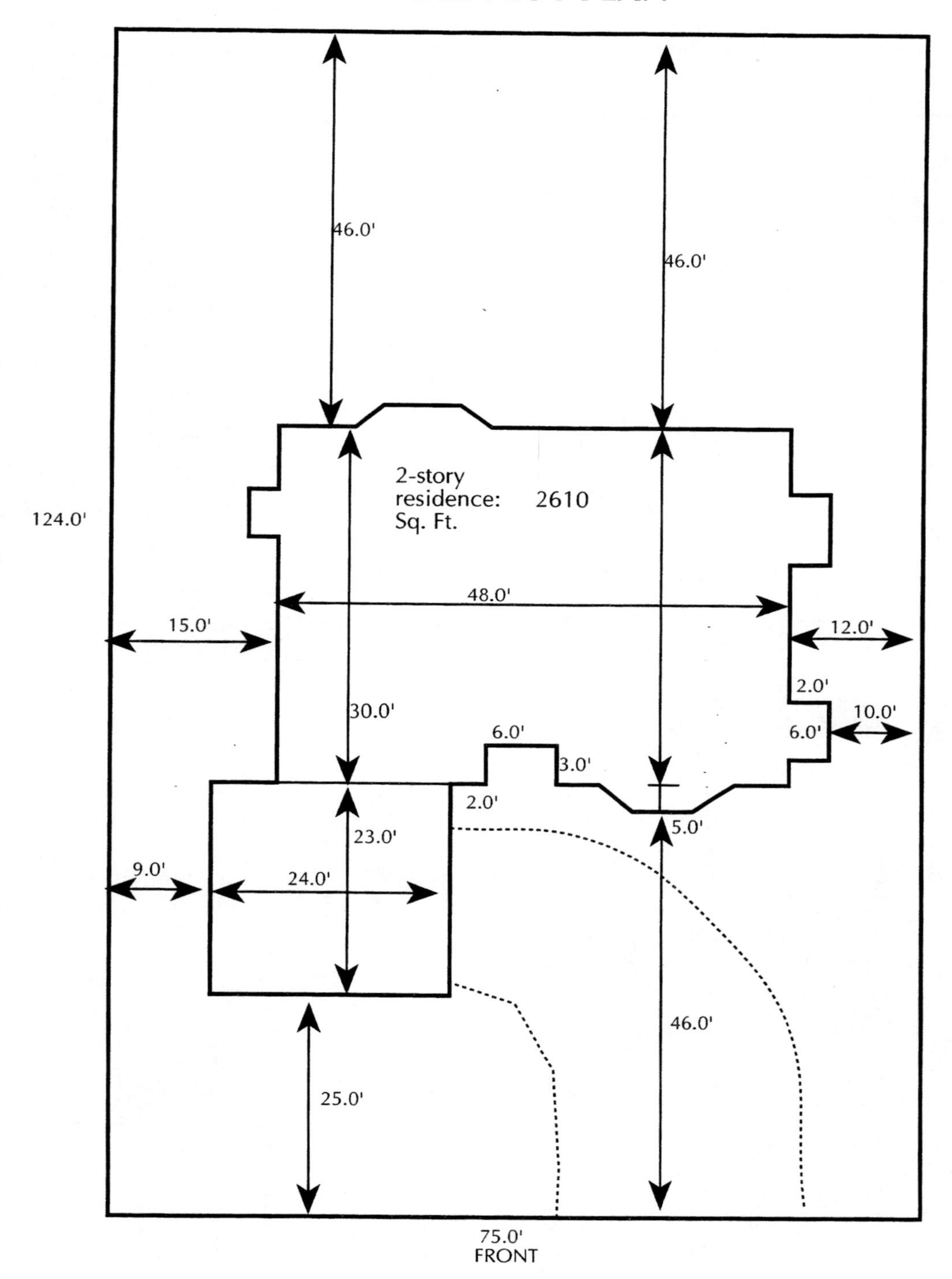

SAMPLE GRADE SLIP

Job No.____________________________ Date____________________ __________
Address____
Lot______________ Block________________ Plan no.____________________
Notify_________________________ Address______________________________

Suggested landscape grade

Top of footing elevation
(see note)
Lowest Top of footing

Note: The top of footing elevation is directly related to the suggested landscape grade. Footings must be of sufficient elevation to allow drainage from the base of the entrance step or landing of the house and from all basement windows when all landscaping has been completed. This elevation cannot be below the lowest top of footing elevation.

Lot grade-Front ______________________
Lot grade-Front ______________________
Lot grade-Rear (at property line) ______________________
Lot grade-Rear (at property line) ______________________
Minimum setback required ______________________
Easement required No_______ Yes_______

Services: Location and elevation inside property line

	Size	Location	Elevation	Weeping tile Yes/No
Sanitary				
Storm				
Water				

Footing elevation certificate required Yes_______ No_______

____
Signature

WORKSHEET # 12

LEGAL AND INTEREST COSTS CALCULATION

1. Legal cost
   (for administering land purchase and registering new mortgage) (approx.) $________

2. Mortgage draw interest
   Interest on progress advances (amount of advance x interest rate ÷ 365 x No. of days paying interest) $________

3. Land interest to developer or realtor, etc.
   (balance owing on land x interest rate ÷ 12 x number of months until it is paid out) $________

4. Interim financing interest
   Approximate average amount required x interest rate ÷ 365 x number of days outstanding $________
   Note. You can only estimate, as the outstanding balance owing will fluctuate daily during construction

5. Credit financing interest
   (usually calculated as 2% per month on overdue accounts to a supplier or trade) $________

   TOTAL LEGAL AND INTEREST COSTS $________

NOTE: If you estimate higher than 3% of your total construction costs (due to a longer than 3 month construction time), enter your estimated figure on your cost summary and loan calculation sheet.

STRATEGY TO REDUCE INTEREST COSTS

1. Build in less than 3 months, 4 months if a larger home over 2000 square feet.
2. Use all lines of credit with suppliers
3. If possible, negotiate an assignment on the mortgage proceeds with suppliers.
4. Consider the lien holdback of 15% for each trade for the lien period.
5. Receive mortgage advances on time to pay down any interim financing which is financed at a higher rate of interest.
6. Use own cash to start construction.
7. Complete all deficiencies as soon as possible to obtain all mortgage monies and pay down all interim loans.

WORKSHEET # 13
CONSTRUCTION SCHEDULE
ACCRUING INTEREST ON ADVANCES
(Based on a 90-day construction schedule)

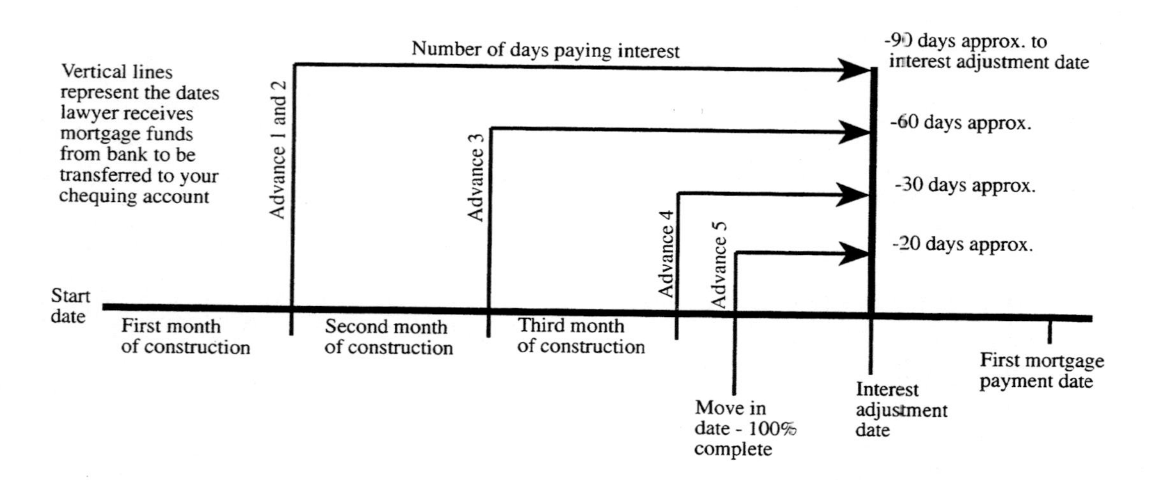

Mortgage Advances	Amount of Advance	Rate of Interest	Days	Amount of Interest
1. Land Advance	$ ________	________ %	90 Days	$ ________
2. Subfloor Advance	$ ________	________ %	90 Days	$ ________
3. Structure and Insulation	$ ________	________ %	60 Days	$ ________
4. 70% Complete	$ ________	________ %	30 Days	$ ________
5. 100% Complete- Move in	$ ________	________ %	20 Days	$ ________

HOW TO USE THIS CHART

Assume advances require one week processing time after inspection. Use the formula to calculate the amount of interest on your mortgage advances for the three month construction period.

Amount x rate of interest ÷ 365 days = interest cost for one day
Interest cost for one day x 90 days = Amount of interest on first advance

The Appraisal Report

When you apply for a mortgage, the lender will have an appraiser forecast the future market value of your property. This is to ensure the lender's security in relation to the loan amount. The lender will charge you for the cost of the appraisal, and it will be up to the lender whether or not to disclose the report to you. Therefore, when you apply for your mortgage, **ask for a copy of the appraisal report.** In some cases, a lender's application fee may be hidden in a $200.00 fee for an appraisal, which costs the bank only $150.00.

Functions of the Lawyer

Even though you pay the legal fees out of your mortgage, the lawyer is acting on behalf of the lender. Depending on the circumstances, you may want to hire your own lawyer; however, it is an additional expense which is usually not required.

The lawyer's functions are:

(a) To search land titles
(b) To transfer the land
(c) To prepare mortgage documents
(d) To obtain receipt of fire insurance policy
(e) To obtain a copy of the survey certificate
(f) To obtain a tax certificate
(g) To execute all documentation with applicants
(h) To forward documentation to land titles office for registration
(i) To report to the lender
(j) To receive mortgage advances and disbursement of funds
(k) Payout of lot
(l) Attend to all deliveries

To do all this, the lawyer needs to communicate with you on several occasions. You should in turn keep well informed on the progress of your advances and the preparation of all the documentation.

Using an Assignment of Mortgage Proceeds

On one occasion the bank would only give me a completion mortgage. They would not approve a draw mortgage or any interim financing to finance the home during construction. I was stuck, I had a mortgage but no financing to build with other than access to about $10,000 from my own cash and available credit. The bank simply could not believe that I could build the home for what I was proposing, thinking it would cost me more!

Fortunately I have a great lawyer who was able to suggest an idea. If I could meet face to face with my suppliers and trades and offer them a legal assignment of the mortgage proceeds as well as agree to their estimate, giving them the work, I might be able to finance the home to completion.

That's exactly what happened, using my $10,000 for some necessary deposits and to pay the concrete cribbers and framing crew (because their contract is all labor and they will need their wages soon after completion), I was able to negotiate up to 90 days credit by offering to them an assignment of mortgage proceeds.

They all accepted the idea because they were guaranteed to get paid (as long as I completed the home which I didn't tell them) the only difference being it will take a little longer and payment would be made through my lawyer once in receipt of the mortgage funds. There is no question, I had to sell them on the benefits of this idea, be accurate with my pricing and build the home as quick as possible.

In the end, the bank was surprised that I built the home and I saved an additional $2,000 in interest that I would otherwise have had to pay the bank for interim financing.

I would suggest offering an assignment of mortgage proceeds to your major suppliers (concrete, lumber, windows) because they are more likely to have the cash resources to allow this extended line of credit. It is not a significant amount of money when interest rates are at a low and would be difficult to negotiate if you were building in boom times.
The exact form I used is illustrated in WORKSHEET # 14.

WORKSHEET # 14

ASSIGNMENT OF MORTGAGE PROCEEDS

TO: ____

____

(name and address of lawyer)

RE: ____

(full names of builders)

Purchase of ____

____

(address of the home to be constructed)

Legal Address ____

and Mortgage to ____

(name of mortgage company)

We, ____

(full names of builders)

do hereby assign, transfer and set over to:

______________________________ (name of supplier)

____

(herinafter called the "Assignee")

the sum of ______________________(__________) dollars, upon receipt of the same in your office for release from the new mortgage ______________________(name of mortgage company)

In witness whereof we have hereunto set our hands and seals this ______ day of __________, 19___.

____

(signature of builder)

____

(witness)

____

(signture of builder,spouse)

The undersigned hereby acknowledges service of this Assignment and agrees to remit funds to the Assignee at the assignee's above address when the funds are received and are available and authorized for release.

______________________________ ______________________________

(name of lawyer) (signature)

Summary Example

The following is an example of a typical applicant who has applied for a mortgage to build his own home. For this example the following information is assumed:

(a) Total cost to construct the improvements plus the land is $150,000 (land $50,000, house $100,000, legal and interest $2,500). The applicant's plan to build a 2,000 square foot two story home for $50.00 per square foot.

(b) The applicant's have $25,000 of their own funds and are applying for a mortgage of $125,000. They used $7,500 (15%) for a deposit on land.

(c) The appraised value of the property is $167,000. Breakdown: $50,000 for land and $117,000 for building. (Appraisal was done by the bank's appraiser using a comparative market approach based on the blueprints and specifications provided).

(d) The applicant and his wife are earning an income of $50,000 combined. His is $25,000 and hers is $25,000 using 100% of both incomes.

(e) They have a side loan of $4,000 with payments of $400 per month.

(f) We also assume the applicant's credit rating is good.

(g) The rate of interest on mortgages is 8%, the corresponding payment factor for a 25 year amortization is 7.63213 (monthly payments)

The first thing to do is qualify the applicants.

Income:	His:	$25,000
	Hers:	$25,000
	TOTAL	$50,000

$50,000 or $4,167 per month x .3 = $1,250 per month.
This represents the 30% allowed for PIT payments.

$1,250 - taxes $250 = $1,000 per month for PI
$1,000 ÷ 7.63213 x 1,000 = 131,025

The applicants would qualify for a mortgage loan of $131,000 but they only need $125,000. Let's look at the TDS and GDS ratios.

$125,000 x 7.63213 (8%) ÷ 1,000 =		954.01
Plus taxes per month		250.00
	PIT	$1,204.01

$$\frac{\text{PIT } 1{,}204}{\text{Monthly income } 4{,}167} \times 1{,}000 = 28.9\% \text{ GDS}$$

PIT	1,204.	per month
Other debts	400.	per month
Total payments =	$1,604.	

$$\frac{1{,}604}{4{,}167} \times 100 = 38\%$$

At this point the applicant's meet all the requirements, the GDS is less than 30% and

TDS less than 40%.

How will this loan be administered? There are several alternatives a bank can consider depending primarily on the loan to value ratio. First, if the loan is above 75% of the future appraised value then it would likely be a high ratio NHA insured mortgage which the banks can administer through CMHC. Second, if the loan is less than 75% and the home is being built, the bank can still request that the loan be insured through CMHC making it a NHA mortgage also. On the other hand, if the loan is below 75% of the appraisal, the bank may hire an independent appraiser to inspect the work in place. The bank would advance funds after acceptance of the inspector's report.

In this case the applicant's are requesting funding for 83% of the total costs and 74.8% of the expected future appraised value (the difference may be considered sweat equity). Some institutions are more agressive than others and will finance up to 75% of the appraised value. Others may finance only 75% of actual costs based on all major items which the builder must provide estimates to validate the amount.

In this example, the loan to value ratio is 74.8%. Since the home is yet to be built, the bank would probably insist that the loan be insured through NHA. Inspections during construction would be carried out by inspector's from local CMHC offices. If the applicant's do not want an insured mortgage and do not want to pay a mortgage insurance fee, they will only be approved if they can find a bank or trust company to lend based on the appraised value (i.e. 75% x $167,000 = 125,250, applicant's need $125,000 mortgage).

In different circumstances, the required loan amount may be less than the construction costs or substantially less than 75% of the appraisal. In this case, the owner- builder may be able to take out a new mortgage on the completed home to pay out the construction loan.

The loan is approved, a commitment letter is drawn up, the forms are signed, and the documents are forwarded to the lawyer to be prepared. At this stage, the applicant should be in a position to start building or relatively close. We will assume construction has started and they are ready for their first advance. They must bring to their lawyer a copy of the survey certificate and proof of insurance. The mortgage documents will be signed by the applicant's, at the lawyer's office, and returned to the lender. The figures would look like this:

TOTAL LOAN AMOUNT		$125,000	
Subtract			
1. Mortgage insurance fee		0	
2. Land advance	$42,500		
SUBTOTAL	$42,500	$42,500	
Progress advances based on		$82,500	
Inspections and advances			
1. Foundation advance (approx. 16%)		$13,200	
Less 15% mechanic's lien holdback (MLH)		1,980	
TOTAL FIRST ADVANCE		$11,220	$11,220
2. Structural advance (approx. 42%)		$34,650	

Less first advance	13,200	
Subtotal	$21,450	
Less 15% MLH	3,217	
TOTAL SECOND ADVANCE	$18,233	$18,233
3. Intermediate advance		
(approximately 70% complete)	$57,750	
Less first and second advances	34,650	
Subtotal	$23,100	
Less 15% MLH	3,465	
TOTAL THIRD ADVANCE	$19,635	$19,635
4. Occupancy advance		
(100% - assume no deficiencies)	$82,500	
Less first, second and third advances	57,750	
Subtotal	24,750	
Less 15% MLH	3,712	
FINAL ADVANCE	$21,038	
Less an amount for accrued interest on previous advances	1,500	
Subtotal	$19,538	
Less legal fees 1,000		
TOTAL FINAL ADVANCE	$18,538	$18,538

Note: Legal fees will be deducted from advances according to the lawyer's payment policies. The mechanic's lien holdback totalling $12,374 on all advances will be released 45 days (no. differs among provinces, check with your lawyer) after the 100% inspection.

Mortgage Summary Breakdown

Total mortgage	$125,000	
Mortgage insurance fee	0	
Land advance	$42,500	
1st advance	$11,220	Foundation
2nd advance	$18,233	Framing + Sub Floor - Pay off Balance of Land
3rd advance	$19,635	
4th advance	$18,538	
Interest deducted	$1,500	
Mechanic's lien holdback	$12,374	
Legal fees	$1,000	
TOTAL	$125,000	

Interest Calculations on Mortgage Advances
(Based on a 3-month construction schedule)

	Advance	Rate of interest	Days	Amount of interest
Land	42,500	8%	90	$838
1st	11,220	8%	90	$221
2nd	18,233	8%	60	$239
3rd	19,635	8%	30	$129
4th	18,538	8%	20	$81
Estimated total interest				**$1,508**

Assume $1,500 interest will be deducted from the final advance.

8

INTERIM FINANCING AND INSURANCE

A. INTERIM FINANCING

Arranging interim financing is arranging for sufficient cash flow. It covers your expenses until mortgage draws are released or until the home is complete and the money spent on construction is converted into a mortgage.

1. WHEN INTERIM FINANCING IS REQUIRED

When you are establishing a line of credit with suppliers, they will telephone your bank to verify that you have arranged for financing. This provides positive proof that you will have the money to pay your account. Unless you make prior arrangements with your bank, you could jeopardize receiving this credit. Remember, credit financing is important for your cash flow and to reduce interest costs.

Even if you feel you do not require interim financing, it is inexpensive to get approval for it from a bank, and there is no interest charged on money you don't borrow. Furthermore, by setting up interim financing you are establishing a line of credit for future borrowing.

2. WHERE TO APPLY

The first place to try is your own bank. Some lenders may refuse to provide you this service unless you are already a customer or plan to become one. Being turned down at one branch does not mean you won't qualify at another, even another branch of the same bank. Many lenders will not have prior experience in administering an interim financing account. If at first you don't succeed, do not give up. Try other banks and credit unions until you find one willing to help.

3. WHEN TO APPLY

Call for an interview after your mortgage is approved and after you have set up adequate fire and construction insurance. When a couple is applying, both must attend at the same time because both signatures will be required. Apply before you begin construction!

4. WHAT TO BRING WITH YOU

(a) A copy of your mortgage approval letter
(b) Proof of insurance
(c) Your cost summary sheet (have all your cost estimates handy)

(d) Personal and financial information as discussed in last chapter
(e) A small amount of cash to open a personal chequing account which will be your house account
(f) Smiles and a positive attitude

5. HOW MUCH INTERIM FINANCING TO APPLY FOR

Assuming your land will be fully paid from mortgage advances, the amount of interim financing you should apply for equals your cost to build less your cash input. You may not require this much, but at least you will have enough approved just in case your mortgage advances are slow and you build fast. You won't have to worry about running short and not completing the home. Besides, you only pay interest on the amount you spend. Study the following example.

Calculating Amount of Interim Financing Needed

Construction costs	$95,000
Contingency factor (2.5% estimated)	2,500
Land	50,000
Total finished home and land	147,500
Legal and interest costs (estimated based on mortgage advances)	2,500
TOTAL COSTS	$150,000

Assume a $25,000 down payment with a land deposit of 15% or $7,500 ($7,500 is part of the $25,000).

Construction costs	$95,000
2.5% contingency factor	2,500
Legal and interest costs	2,500
Total costs to build the home	$100,000
Subtract the amount of cash input by builder (25,000 - 7,500 = 17,500)	-17,500
Total amount of interim financing the builder should apply for	$82,500

In this example, the builder would apply for enough financing to build the home to completion without having to rely on mortgage advances. Those builders who will be searching for a mortgage after completion will require enough interim to complete the entire project.

6. HOW THE MONEY IS ADVANCED

Advances will be determined by bank policy, usually in increments from $1,000. to $5,000. (The ideal situation is to have an operating line or credit where the bank will deposit only what you require.) Call the bank when you first begin to write cheques and advise them to deposit money into your account. Keep in touch with your loans officer.

The advances should be immediately applied to any overdraft in your account with the remainder being applied against your loan.

EXAMPLE OF LENDING PROCESS

Cheques Drawn On Your Account	Date	Bank Balance	Interim Loan(s)	Total Loan
	1st	$5,500.00 Starting	------	
$1,000.00	2nd	$4,500.00	------	
$2,500.00	4th	$2,000.00		
$2,000.00	8th	$ 0	Call Bank	
$1,000.00	10th	$1,000.00- Overdraft		
$1,500.00	15th	$ 500.00	+ 3,000.	$3,000.00
$2,000.00	20th	$1,500.00	+ 3,000.	$6,000.00
$2,000.00	25th	$ 500.00- Overdraft		$6,000.00
$1,500.00	30th	$1,000.00	+ 3,000.	$9,000.00

The above example is showing funds advanced into your house account in increments of $3,000. each.

7. HOW MUCH IT WILL COST

You pay interest only on the amount you receive. The rate will be based on current bank prime rates. A normal agreement for interim is prime plus three. There may be a registration fee of $100 to $500, depending on the bank. It may be possible to obtain a lower rate, say prime plus one; however, this depends on the applicant's credibility.

8. HOW THE INTEREST IS PAID

The interest is usually added on to the loan each month, which means after the first month you will be paying interest on your interest. Keep track of the interest costs each month and be sure to add them into your chequebook. Your balance must coincide with that of your banker.

9. KEEP A RECORD AND BALANCE OF YOUR ACCOUNT

You are your own bookkeeper and therefore are responsible for balancing your account and keeping track of the money you spend building your house. When two parties (husband and wife) will be writing cheques, it is extremely important that you communicate and use only one book for balancing. Each month, record the service charges, loan interest, and overdraft interest in your house account chequebook.

It is important to keep a separate record comparing your costs to your original budget. See Chapter 13 on recording and cost control.

If you lose track, go to your bank and find out what cheques have gone through your account. Call your lawyer to check on the mortgage funds advanced to date. Find out where you stand financially by comparing the bills and the interim loan you have to pay with the mortgage advances still to come. The worst thing to do is to go on building and assuming you have plenty of money!

10. LENGTH OF THE LOAN

The loan period depends on how long it will take you to build. Check with your lender during the interview. Interim financing loans are usually set up as "demand loans" due in full on a certain date. Since some of your mortgage money won't be released until after the home is complete, the interim loan period must be for longer than your estimated construction period (e.g., 3-month construction period = 5-month interim loan period).

If you go over your allowable budget, go to your bank and ask for help. Explain why you went over and how much additional you require. If the amount is small, you might arrange for an overdraft or a small personal loan with no security. If the amount is large (i.e., over $5,000.), the bank will want security in the form of a second mortgage provided you have the ability to make payments. The solution is simple: DON'T OVERSPEND!

11. HOW LONG DOES IT TAKE TO GET APPROVED

You should be able to get either a "maybe" or a "no" answer during your interview. It will take a couple of days for the bank to verify your credit and give you a formal yes or no.

12. WHAT SECURITY WILL THE BANK REQUIRE

Interim financing loans have relatively little security. The bank will obtain your personal "I promise to pay" on their notes. Furthermore, the bank will insist on obtaining a general assignment of all the mortgage proceeds. The assignment will guarantee the interim loan provided you spend the money on the home and the home is properly built. An example of an assignment or letter of direction is shown below.

Letter of Direction

To: *Name of mortgage lenders solicitor*

You are hereby authorized and directed to pay from the proceeds of a mortgage made between *The mortgage lender* as mortgagee and *Names of applicants* as Mortgagor(s) the sum of *Amount of interim financing* ($) Dollars to the *Interim lender and address*

as well as interest thereon, from the date of advancement of funds up to and including the date on which the funds are paid in full to the said bank, at the rate of interest as set out by the bank.

We further covenant and agree that if, for any reason, the full proceeds of the *Mortgage lender* Mortgage are not advanced after registration of the lands in (my) name(s), the said sum of *Amount of interim financing* ($) Dollars shall constitute a charge on the following lands until paid:

Lot ______
Block ______
Plan Number ______

We further agree and confirm that we have placed adequate and proper fire insurance coverage on the subject lands and premises to adequately protect the mortgage company.

This authority is irrevocable and cannot be withdrawn without the consent of *The interim lender*

Dated at *Name of town or city*
in the province or state of ______
this ____ day of ______
19____

Name of applicant
Name of co-applicant or spouse

Bank officer
Witness

Bank officer
Witness

B. YOUR INSURANCE REQUIREMENTS

Every homeowner who has a mortgage will be required by the lender to carry property insurance for the usual hazards of fire, windstorm, hail. While under construction, you will need additional protection against *vandalism, theft, and glass breakage.*

No one can hedge against every possibility of loss. Every person must analyse the situation, calculate the risks and the cost of covering each risk by insurance, and choose accordingly. Your insurance coverage should meet well-defined objectives. The policies, like your clothes, must fit you. The following discussion will give you some insight about the coverage you may require.

1. WHEN TO OBTAIN INSURANCE

Arrange insurance before you begin to build! As soon as you excavate, you have a risk of liability. As soon as materials are delivered, you have a risk of theft. If you are seeking financing, you will have no choice. Every lender will demand proof of insurance before advancing funds.

2. THE INFORMATION REQUIRED

The insurance company will require your present and future civic mailing addresses and the legal address of the lot (i.e., Lot, Block, Plan Number). In addition, they need to know the minimum amount of insurance your lender requires you to maintain. This information should be on your mortgage approval letter, so simply bring a coy with you when you apply.

3. HOW MUCH IT COSTS

Shop around and compare premiums. Fire insurance premiums will be similar, but vandalism, theft, and glass insurance while under construction can differ substantially between insurance companies. Find out how much deductible is charged and if you can get a lower deductible for an extra charge. The amount of the deductible is the amount the insurance company will deduct from the value of the loss to arrive at a settlement, i.e., with a $1,000. loss, the owner pays the first $250.

4. LIABILITY INSURANCE REQUIRED

Whether building or not, your basic fire insurance should include protection against liability hazards. This insurance will indemnify you if you are legally required to pay damages for bodily injury or property damage arising out of your premises or personal acts.

5. COVERAGE NEEDED FOR CONTENTS

The basic fire insurance package will include a portion for furnishings and personal effects. The amount of coverage is frequently expressed as a percentage of your home coverage (e.g., 60% of building amount). Check with your agent to ensure your particular needs are being met.

In a loss, an insurance company may pay you the depreciated value of your furnishings and personal effects. For a little extra money you can obtain replacement value on your contents. This is valuable protection, especially when inflation is increasing those replacement costs. Your house insurance should automatically keep up with inflation.

Check with your agent about coverage for tools used during construction. Your own tools should be covered, but usually not tools left by other trades.

6. MORTGAGE INSURANCE

Unless taken out separately, your regular home package will not include mortgage insurance. Mortgage insurance is decreasing term life insurance that provides ample funds to pay the house off in case something happens. A husband and wife can cross-insure themselves to protect their mortgage. Similar to life insurance, the younger you are the cheaper it is.

7. INSURANCE TIPS

- Place a sign on your property, for example, DANGER--KEEP OUT! TRESPASSERS WILL BE PROSECUTED, REWARD FOR REPORTING VANDALISM AND MALICIOUS ACTS.
- Make a list of your personal possessions and keep it in a safe deposit box or other safe location.
- If you have a theft on your property, get a police report as soon as possible.
- Have your windows delivered just prior to installation.
- Install a lock as soon as possible once your home is closed in.
- Backfill trenches after inspections have passed.
- Always think safety. When building a temporary walkway, make it safe for *anyone* to walk on.
- Keep receipts for everything you purchase.
- Check the possibility of obtaining worker's compensation if you and your friends will be doing considerable work.
- Advise insurance agent immediately upon possession of your new home.

9

House Placement and Grading

House Placement

As a general rule, you cannot make a decision on house plans until you have purchased your lot. When you have selected a lot, you need to find out the zoning regulations pertaining to the setbacks from the street or lane and the minimum side yards. This information can be obtained from the land developer or from the by-laws of the town office or city planning department.

Use Worksheet #15 on house placement regulations when obtaining local zoning information.

Fit the house to the land and the land to the house. The size and grade of the land can determine the size and type of house you build. Pie-shaped or cul-de-sac lots tend to be suited for a design that will allow the house to be placed closer to the street and nearer to the minimum setback requirements. Some examples include front to back splits, two story designs, and narrow bungalows or bi-levels with possibly a jog in the wall or a front drive garage. The purpose is to cut down on wasted yard space by not setting the house back too far on the lot and remain in conformity.

Side yard space is usually wasted space, so you will want to build close to the maximum width of your lot. By doing so you will gain maximum usable space in the back yard. Also, if your home is wider, more can be done architecturally to enhance the front appearance of your home, which can increase the market value.

It's a good idea to plan in advance for any future developments on your site (additions, garages, decks, future fireplace, and accessory buildings). Even something as small as a fireplace chase can infringe upon the minimum side yard requirements and prevent future development.

Before you buy, visualize how you intend to make the architecture suit the land. How important is passive solar energy to you? South exposure of windows will be the most efficient in producing passive solar energy gains. Try to reduce the number of windows facing north. Furthermore, by strategically placing trees and fences you can provide some protection and shade from sun, noise, drifting snow, and prevailing north winds. The following plot plans show how pre-planning can be of benefit.

PLOT PLAN A

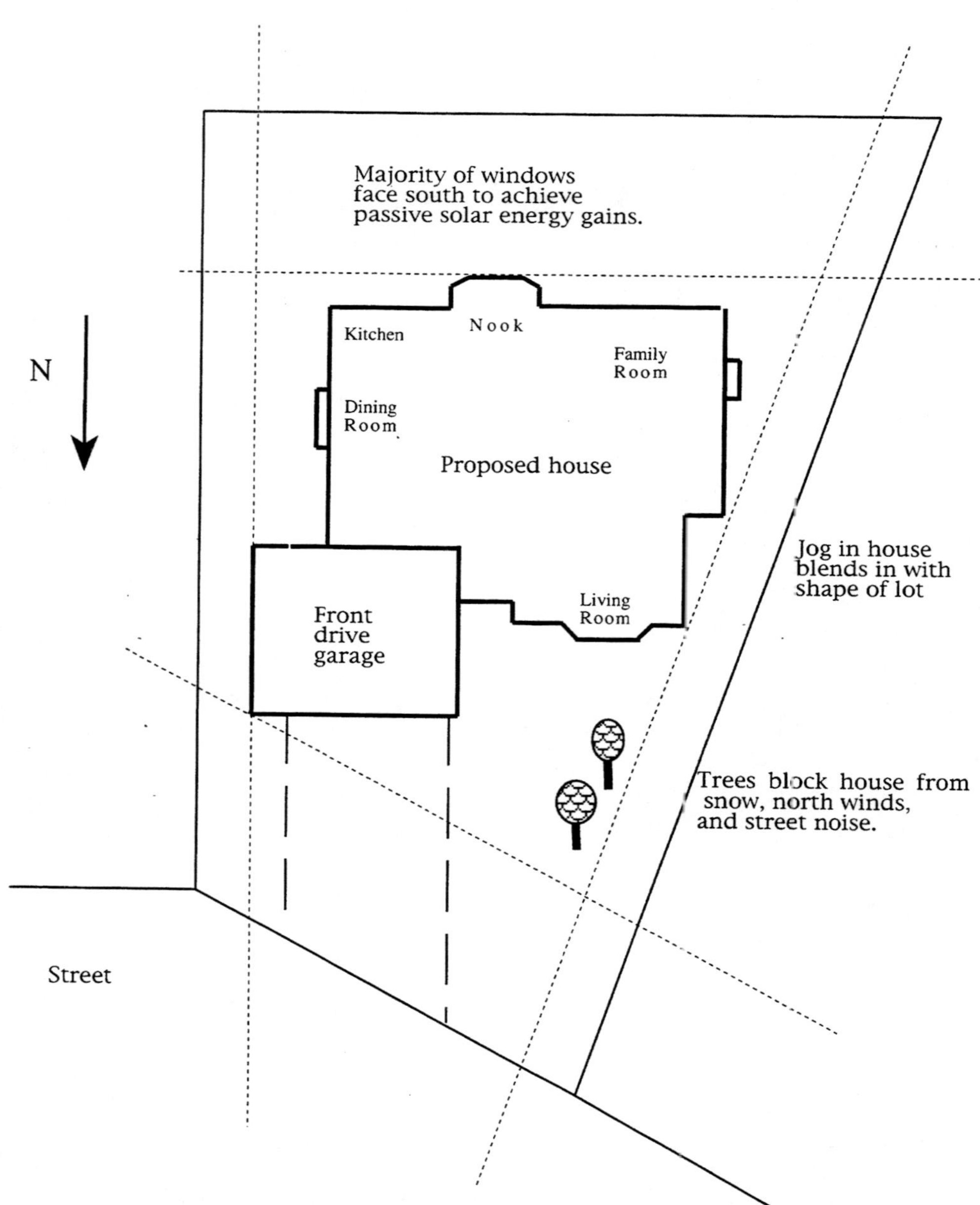

Plot Plan A illustrates the placement of a house on a lot. The dotted lines represent the required sideyards, setbacks and rear yard. The space allowed for construction is the space between these dotted lines. When custom designing a home to make the best use of the available construction area, the draftsman should first identify the available space for construction.

PLOT PLAN B

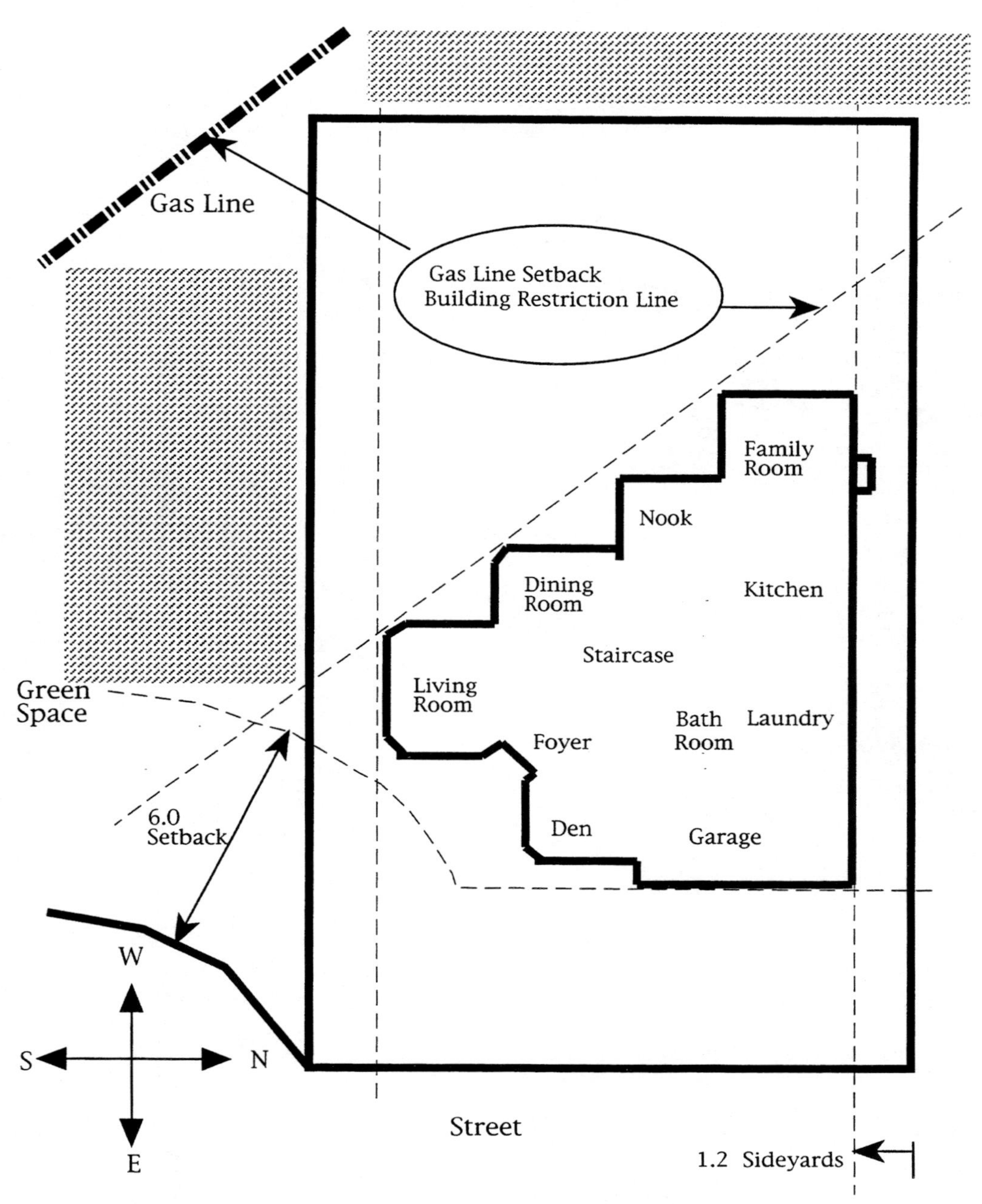

Plot Plan B shows a building lot with several restrictions requiring a creative custom design. By jogging the home along the building restriction line the builder was able to gain the benefits of well defined rooms which face south and west.

WORKSHEET #15

HOUSE PLACEMENT REGULATIONS

Interior Lots

	WITH LANE	WITHOUT LANE
Setbacks:		
FRONT	________	________
REAR	________	________
SIDE YARDS	________	________
DRIVEWAY REQUIREMENTS	________	________

Corner Lots (house facing short side)

Setbacks:		
FRONT	________	________
REAR	________	________
SIDE YARDS	________	________
DRIVEWAY REQUIREMENTS	________	________

Other Regulations (maximum site coverage, fence height, deck, carport)

Other ____________________

Garage Placement Regulations

Maximum or minimum size	________	________
Setbacks:		
FRONT	________	________
REAR (with front drive)	________	________
LANE (with side drive)	________	________
LANE (with rear drive)	________	________
HOUSE (if detached)	________	________

Side yards:

Garage located in back yard ____________________

Garage located in front yard ____________________

GRADING

In a controlled subdivision it is critically important that the grade of your lot conform to the final grades given to you by the developer or city planning department. A mistake here can cost you a bundle in market value! Your home will look out of place if it sits up too high. On the other hand, you won't receive proper drainage if the home is too low.

Improper grading can affect the drainage on your neighbours property. In this case, the person at fault has to construct a retaining wall or change the grade to conform to the neighbours house. To be safe, have the elevation checked after you excavate, and plan in advance the drainage on your lot.

The elevation check is usually done by the engineer controlling the development, however, if there is no land developer requesting this inspection, it is your option to perform this inspection as a safety check. Your surveyor has the equipment and is capable of doing the job.

The diagrams on the following page illustrate examples of property grading.

Lot showing split drainage

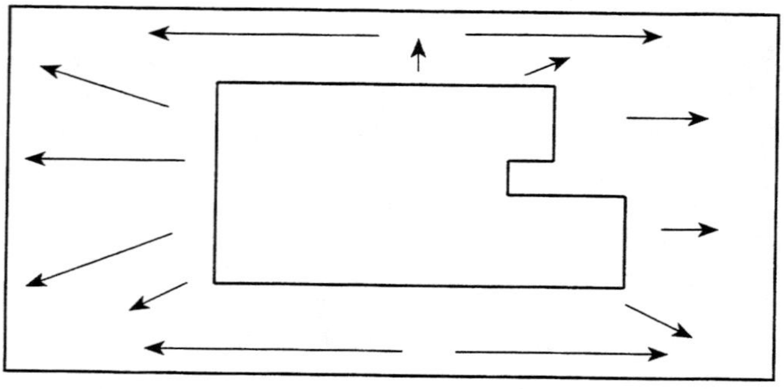

Arrows show direction of water flowing to the back and to the front of the lot.

Cross section of grades

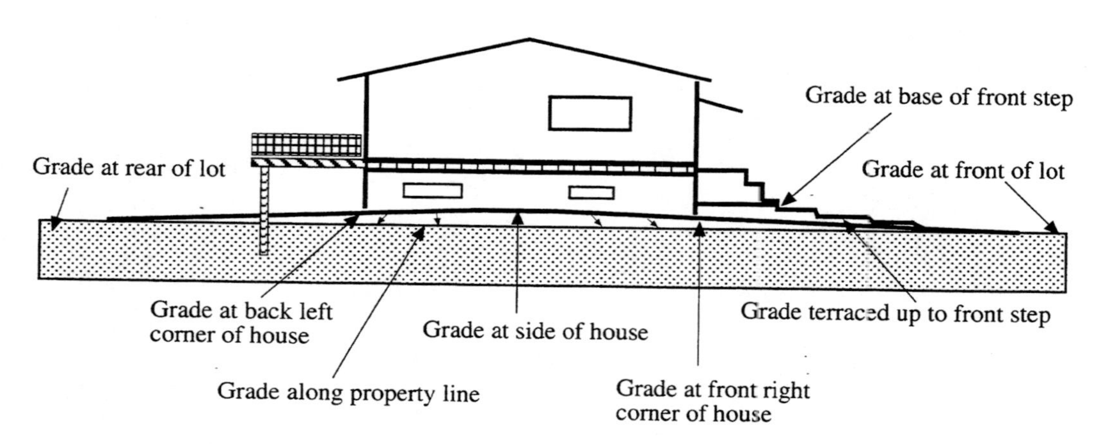

Careful attention must be given to all the required grades when building in a controlled subdivision. Check the grades on the four corners of the house, the sideyards and the four corners of the lot. Most grades are marked on a grade slip provided to you by the developer or their engineers.

Drainage is always away from house, even when the lot has a steep slope.

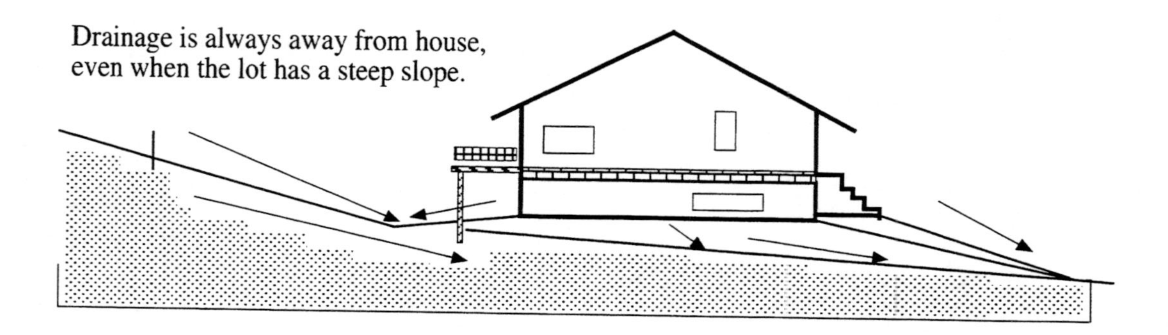

10

HOUSE DESIGN AND PLANNING

APPRAISING YOUR DESIGN - STRATEGY

Pick a good house design and plan well any innovative changes. This is the key to success and savings. Since the money you save will be determined by the *future market value* of the home, your first task in choosing a design is to decide what criteria you must meet in order to realize the highest potential value in the marketplace.

Conduct your own appraisal by considering the following:

- Does the style of home conform with others in the area (traditional, ranch house, contemporary, etc.)?
- Does the size of home conform with others in the neighbourhood?
- Does the home have an efficient floor plan that suits your needs?
- Will the home be passive solar energy efficient?
- Is the design original?
- Will it be in harmony with its exterior surroundings: sun, wind, topography, drainage, view, privacy, street access?
- Will the plan make use of products that will contribute to value?
- Is it affordable?
- Will it meet the architectural guidelines, if any, for the area?

Analyse the above points. Will the plan suit your needs and budget?

After compiling your initial estimated figures for house and land, you may discover that it does not fit your financing. Your next step is to begin making the appropriate budget trade-offs and adjustments to your plan. Consider doing one or more of the following:

- Change the type of house
- Adjust house size, shape
- Save money by doing more work yourself
- Limit the number of extra items, plan others for future
- Control quality and costs of carpets, cabinets, fixtures, etc.
- Purchase a more affordable lot
- Alter the construction materials to fit budget (brick, studding, insulation, foundation, etc.)

COMPARING HOUSE TYPES

Which house is for you? The least expensive is a small rectangular one-story building, such as a bungalow or bi-level. A two-story house with the same size foundation and roof gives double the living area, but at less than double the costs. A split level of the same original foundation size increases the living area over the one-story but at a substantially higher cost.

Each basic house type has its own definite advantages and disadvantages. These should be analysed carefully prior to making your selection.

BUNGALOW

ADVANTAGES

- Easiest and fastest to build
- Simplicity means low cost production methods
- Low front profile can enhance front appearance
- Extended wall, garage or car-port can give impression of a larger, wider home
- Allows a good opportunity for floor planning
- Eliminates stair climbing
- Easy maintenance

DISADVANTAGES

- Small basement windows make this area less attractive for future development
- Larger floor plans are more costly per square foot than a two-story design (you must double the roof and foundation area)
- Difficult to heat if plan is too spread out
- Lot may be too small for a one-level design

BUNGALOW

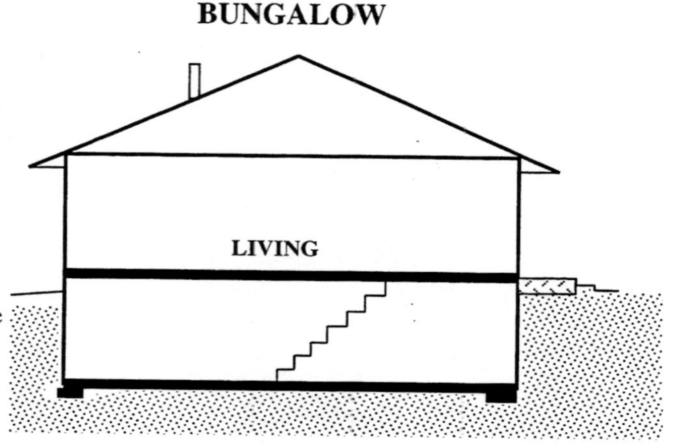

BI-LEVEL

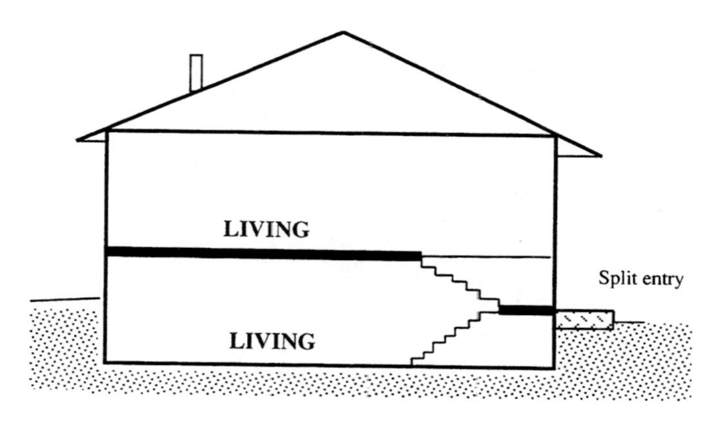

BI-LEVEL

ADVANTAGES

- Split entry allows easy access to both floors
- Less excavation than a bungalow - cost savings
- Larger basement windows make it more attractive for future basement development
- Cost savings on the foundation by using a 4-foot concrete wall or wood foundation
- Only one level and less detail in roof design make it easier to construct than most two-story or split level designs

DISADVANTAGES

- Higher out of the ground than a bungalow and subsequently requires a design with plenty of break-up to prevent a box appearance
- Requires more siding and windows than a bungalow design - cost disadvantage
- In some areas this design may be too common, limiting future resale value
- To be unique requires plenty of research and planning

ONE AND ONE-HALF STORY

ADVANTAGES

- Uses less ground area than a bungalow or bi-level
- A good choice for a small lot or zero lot line development
- Possible cost savings from smaller foundation and roof
- May be easier to heat
- Second story can easily be planned for future development

DISADVANTAGES

- Appearance of steep pitch roof may damage market value if not well designed to blend in with

other homes and the site
- Small basement windows are less attractive for future basement development
- For some people extra stairways may be a disadvantage

TWO-STORY

ADVANTAGES

- Good design for cathedral or high vaulted ceilings up to the second story
- The upper bedrooms have more privacy
- More efficient to heat than a bungalow or bi-level
- Uses less land area making it a good choice for a narrow lot insufficient to accommodate a bungalow
- Lower costs of foundation and roof - cost/square foot is generally less with larger homes
- Easy to separate living, eating, and sleeping areas

DISADVANTAGES

- Must offset the rectangular appearance by planning break-up to the high walls using different roof lines and exterior finishes
- More difficult to build than a bungalow or bi-level
- May be harder to clean
- Additional costs - stairs, framing, some extra siding and windows
- Should have at least one bath on each floor

ONE AND ONE-HALF STORY

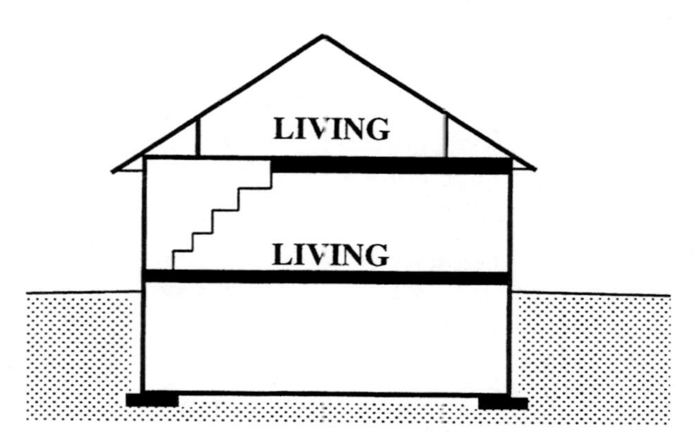

TWO STORY

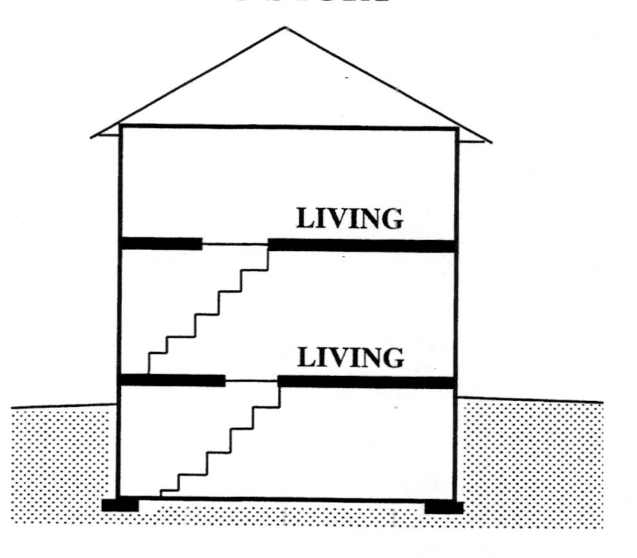

FOUR LEVEL SPLIT

ADVANTAGES

- Easy separation of living, eating and sleeping areas
- Split levels and two-story splits are eye appealing and have good resale value
- Can be designed to suit a level or sloped site
- Allows flexibility by initially developing either 2, 3 or 4 levels
- Flights of stairs are shorter
- Highly adaptable to many styles and ingenious arrangements of roof lines and adjustments of levels

DISADVANTAGES

- Split level homes are the most costly to build - extra material and labour is required in the roof, stairs, subfloor, and finishing
- More stair climbing than any other design
- Often more difficult to landscape
- More land is required than for a two-story, but more liveable space is possible on the same land than in a bungalow

FOUR LEVEL SPLIT

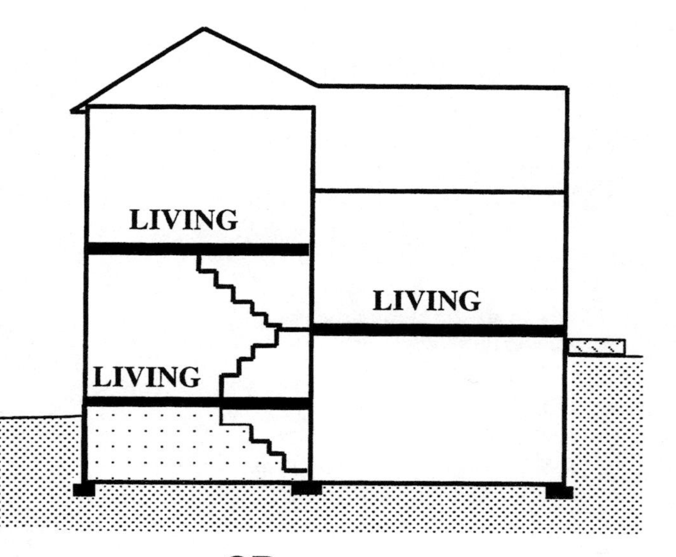

OR

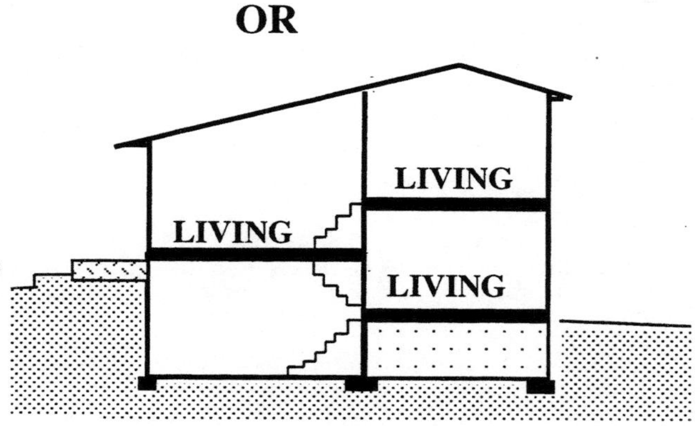

THREE-LEVEL SPLIT

A three-level split can be made by eliminating the fourth level and designing it for a crawl space.

ADVANTAGES

- Provides easy separation of living, eating, and sleeping areas
- Less expensive than a four-level split
- Lower profile than two-story (roof slopes to second story)
- Can easily develop future living area on third level
- Can have an attractive front elevation
- Split levels are well suited for sloped lots; however, can be built on any site (sloped lot may be suited for a grade level entry off the third level)

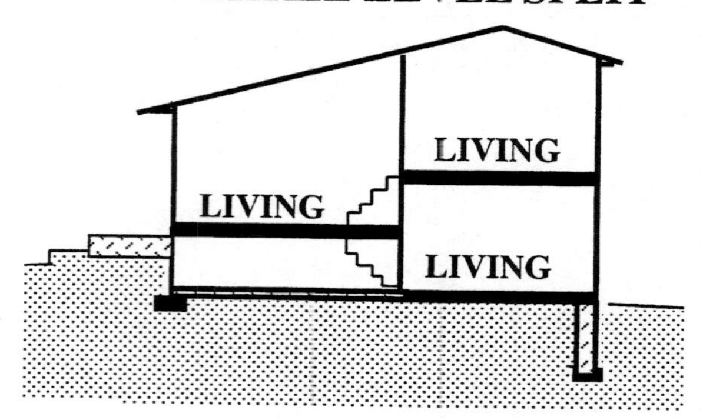

DISADVANTAGES

- Little space for storage and no basement space for future development
- Little difference in cost between a three and four-level split
- Resale value can be limited because potential buyers might want four levels

ACTIVE SOLAR HOMES

Homes designed specifically to accommodate active solar heating are not very practical and for this reason not common. A look at some of the disadvantages will help to explain why.

ADVANTAGES

- Fuel savings

DISADVANTAGES

- May take the life of the system just to recover its cost
- System must be maintained - upkeep costs
- Design is based on saving energy, which could take away from appearance
- Similar fuel savings can be achieved at a lower initial cost with passive solar designs

PASSIVE SOLAR DESIGNS FOR ANY HOUSE TYPE

ADVANTAGES

- Easy for owner-builder to plan
- Little additional cost of sealing the home tight
- Some designs have received up to 85% of heat from the sun
- Life of the system is the life of the house
- Little or no maintenance problems

DISADVANTAGES

- May be difficult to control heat in summer
- Increased cost over conventional - amount depends on design and quality of products
- Most windows must face south
- Built primarily by owner-builder and some small contractors

PASSIVE SOLAR DESIGNS FOR ANY HOUSE TYPE

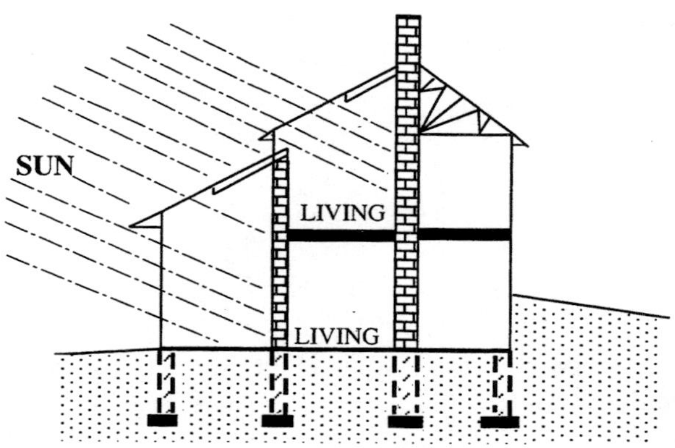

- Majority of windows face south
- Mass wall collects suns heat
- Larger south soffit (overhang) prevents summer overheating
- Some aspects are adaptable to many different designs
- Minimum of R-20 insulation in walls, R-40 ceiling with a 6 ml vapour barrier
- Higher backfill on north side
- Well insulated basement

BASEMENTLESS

ADVANTAGES

- Savings in construction costs. A concrete slab on grade costs less than a plywood and joist subfloor on a full foundation
- Used mostly for modular or mobile homes and on certain lots where a full foundation is not feasible.

DISADVANTAGES

- Special attention must be given to ensure adequate protection from frost, moisture, and cold.
- A basementless house still requires footings and frost walls or piles.
- No basement - no future development

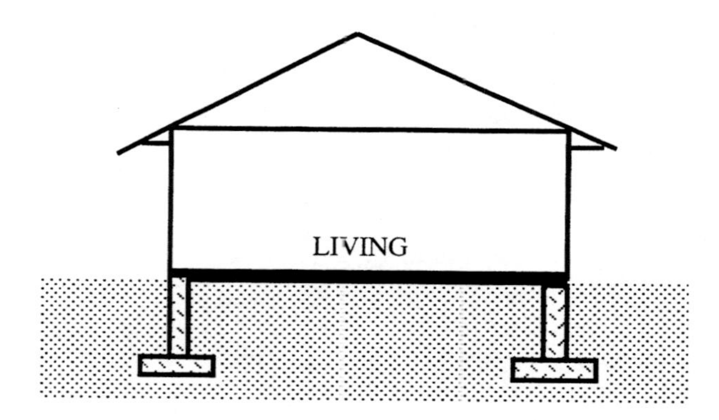

COMPARING HOUSE SHAPES

There can be additional costs stemming from the basic shape of your house or floor plan. Consider the following two alternatives.

The best house shape is economical and also attractive. An example is a rectangular home with a third gable on the front to enhance its appearance. The home should look warm and inviting from the street. Such curb appeal is very important for resale value and your own personal pride. Analyse your proposed design for the strengths and weaknesses of its shape and make the appropriate trade-off decisions.

COMPARING HOUSE SIZES

It is obvious that the larger the home the more expensive it is to build. However, within certain limits, the larger the home, the less the cost per square foot. Consider the following examples, assuming the same house is on identical lots. House #2 has a third bedroom on the main level (10 x 10). House #3 has a third bedroom and a dining room (each 10 x 10).

COMPARING HOUSE SHAPES

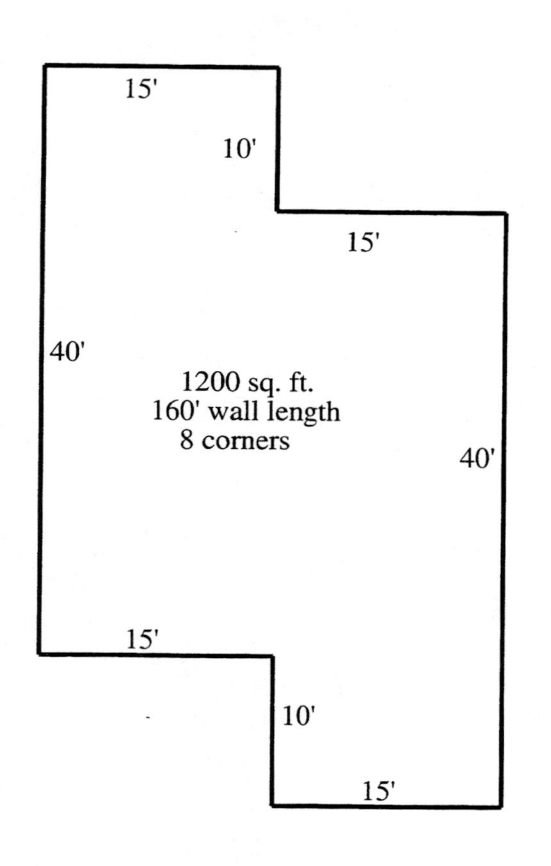

- More expensive design shape
- Has 4 additional corners and 20 extra feet of exterior walls
- Additional gable will add to front elevation and appearance of home.

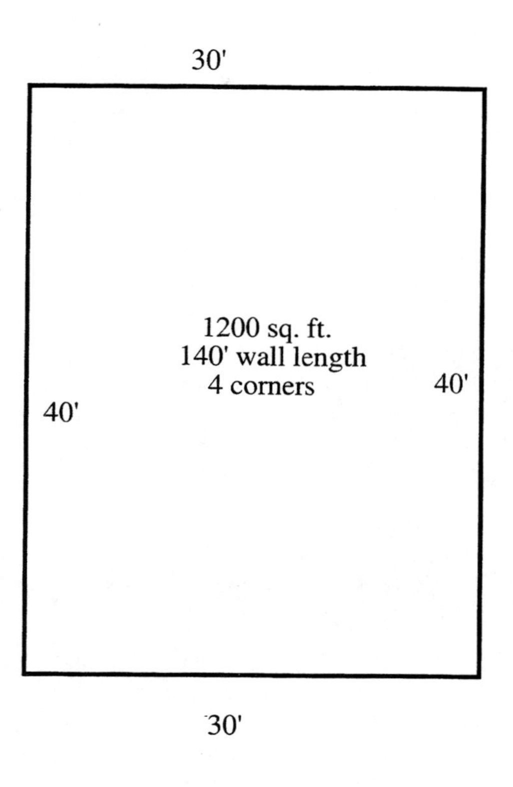

- Economical design shape
- Easiest to build
- Tends to look too boxy

COMPARING HOUSE SIZES

#1

1,000
Square feet
Two bedrooms

#2

1,100
Square Feet
Three bedrooms

#3

1,200
Square feet
Three bedrooms
and dining room

SUMMARY

- All three homes have identical costs for kitchen, plumbing, steps, sidewalks and survey.
- All three homes have similar costs for heating, electrical, excavation, framing labour, interior finishing, services and permits, and your management time required to build each.
- House #2 and #3 have extra costs for basement, framing materials, windows, siding, drywall, and carpets.

When the house size is increased, not all the costs of construction increase in the same proportion. House #3 will have the highest total cost but the lowest cost per square foot to build. Now, what are the potential appraisal or market values of these homes? Suppose the going rate is $50 per square foot. Without including the land, house #1 will be worth $50,000 - house #2 will be worth $55,000 - and house #3 will be worth $60,000. The cost for the extra bedroom and extra dining room in this example is substantially less than the potential gain in market value.

Important trade-off decisions must be made regarding house size. Within certain limits, you are better to build a larger home; however, don't get caught building one you cannot afford. Once you go beyond a certain size, you will be increasing the size of the kitchen, the size of the furnace, the number of bathrooms, and so on. At this point the cost per square foot will increase, and you will be working under a whole new set of guidelines.

HOUSE PLANNING TRADE-OFFS

Your home may represent the largest investment you will make in your lifetime. There is no substitute for good planning, good materials, and good workmanship. In the planning stages, you have a tremendous number of trade-offs to consider.

Begin your planning by studying the house planning worksheet. Check off the design features that are important for your satisfaction and needs. Those items you mark as having high importance should be specific; they will comprise the list of everything you plan to do in your proposed new home. Only by planning well in advance will you be satisfied with the end result.

WORKSHEET #16

HOUSE PLANNING

	Importance	
	High	Low
Requirements of occupants		
Provision for guests	________	________
Provision for others (in-laws, etc.)	________	________
Pets	________	________
Profession of owner (office in home)	________	________
Individual requirements		
Formal entertaining	________	________
Separate formal dining area	________	________
Informal living areas	________	________
Outdoor living and eating areas, decks, patios	________	________
Supervised outdoor play area	________	________
Nursery	________	________
Recreation areas	________	________
Hobby areas (music, sewing, gardening, etc.)	________	________
Study, reading area, office	________	________
Laundry area	________	________
Screened porch or breezeway	________	________
General design		
Bungalow	________	________
Bi-level	________	________
One and one-half story	________	________
Two story	________	________
Split level	________	________
Crawl space	________	________
Concrete slab	________	________
Full basement	________	________
Number of exterior entrances	________	________
Traditional exterior	________	________
Colonial French Tudor	________	________
Country exterior, farmhouse	________	________
Exterior having a variety of different styles	________	________

Roof styles

Gable, split gable, third gable, continuous gable	________	________
Hip roof	________	________
Shed roof, double shed	________	________
Flat roof	________	________
Tudor design, cottage roof	________	________
Dutch gable	________	________

Dress-up ideas

Trim around windows	________	________
Cedar shakes in the gable end	________	________
Lattice work	________	________
Exterior light fixtures	________	________
Brick work	________	________
Combining stucco with siding	________	________
Shutters on front windows	________	________
Window grills	________	________
Vertical cedar siding below windows, etc.	________	________
Batten boards	________	________
Gable dormers	________	________
Pillars	________	________
Mail box and house numbers	________	________
Attic vents located in third gable	________	________
Front porch	________	________

Extra items - budget restriction

Quality of interior trim	________	________
Quality of plumbing and light fixtures	________	________
Quality of floor coverings	________	________
Quality of cabinets, etc.	________	________
Fireplace, woodstove	________	________
Jacuzzi, bar	________	________
Sunken living room	________	________
Half bath or rough-in plumbing	________	________
Bow or bay window, casement windows, skylight	________	________
Patio door and deck	________	________
Vaulted, open beam, sloped or cathedral ceiling	________	________
Laundry room / shoot	________	________
Cold storage in basement	________	________
Garage	________	________

Extra items - budget restriction

Concrete driveway	________	________
Built-in vacuum system	________	________
Basement entrance or third level entry	________	________
Built-in appliance(s)	________	________
Built-in china cabinet	________	________
Feature wall	________	________
Fenced yard	________	________
One piece molded bathtub	________	________

Mechanical equipment required

Forced air heating	________	________
Hot Water	________	________
Electrical	________	________
Solar	________	________
Air to air heat exchanger	________	________
Power humidifier	________	________
Air conditioning	________	________
Intercom system	________	________
Security system	________	________
Washer/dryer	________	________
Freezer	________	________
Exhaust fans	________	________
Water softener	________	________
Size of hot water heater	________	________
220 electrical outlets	________	________

Storage areas required

Cold storage	________	________
Entrance closet	________	________
Bedroom closets	________	________
Linen closets	________	________
Toy storage	________	________
Kitchen equipment storage	________	________
Cleaning equipment storage	________	________
Tool storage	________	________
Gardening equipment storage	________	________
Hobby equipment storage	________	________
Other	________	________

Kitchen requirements

U-Shape	________	________
L-Shape	________	________
Galley shape	________	________
Snack bar	________	________
Built-in oven	________	________
Built-in microwave	________	________
Double sink	________	________
Built-in dishwasher	________	________

Dining requirements

Separate dining room	________	________
Kitchen nook	________	________
Included in living or family room	________	________
China cabinet location	________	________
Table size	________	________

Living room requirements

Location of window	________	________
Location for piano, organ	________	________
Book storage	________	________
Fireplace	________	________

Family room requirements

In basement	________	________
Included on main floor	________	________
Fireplace location	________	________
TV, stereo location	________	________
Other	________	________

Bathroom requirements

Located on main floor	________	________
Located near back door	________	________
Number of sinks	________	________
Separate shower	________	________

Utility room requirements

Located in basement	________	________
Located on main floor	________	________
Utility tub	________	________
Freezer	________	________
Shelving	________	________
Ironing area	________	________

Sewing area ________ ________

Bedroom requirements

Room for king size bed ________ ________
Separate closets ________ ________
Study desks ________ ________
Bunk beds ________ ________

Den/Office requirements

Built-in shelving ________ ________
Room for desk, filing cabinet ________ ________

SELECTING YOUR FLOOR PLAN

Before you begin looking at house plans, you need to understand how to tell a good floor plan from a poor one. An efficient floor plan will separate working, living, and sleeping areas. Traffic patterns will allow easy access from one room to the next without having to cross through rooms in the process. The rooms will be functionally designed with respect to others and minimize wasted hallway space.

In the planning stage, try not to worry about precise details and do not commit yourself to a single idea too soon. You and your architect can incorporate the fine details after you select the basic floor plan layout.

WHERE TO LOOK FOR PLANS

One of the keys to success for the owner-builder is finding a good house plan. You need to look at hundreds of house plans, pick out a few alternatives, and make a decision after analysing each alternative. Some places to look are:

- Your public library is an excellent resource centre (Check under house construction and architecture.)
- House design magazines on newsstands
- House design books in bookstores
- Government books on low energy house designs
- House design services in the phone directory
- Architects, Draftsman
- Check real estate section of weekly newspaper for advertisements of house design catalogues
- Lumber retailers
- Package home builders
- Existing homes - purchase plan from owner
- Builders' floor plans on 8 x 11 sheets - cannot copy identical plan

In the majority of cases, you will want to have your own plans drafted up or make changes to an existing plan to suit your family needs. In either case, your next step is to find a good architect.

IMPORTANCE OF FINDING A GOOD ARCHITECT/ ARCHITECTURAL DRAFTSPERSON

Good drawings will be a big asset when it comes to estimating prices and construction. Too often this job is hired out to the first available person who has some drafting experience or to the person who is the least expensive. In many cases, the builder regrets the decision because of problems arising from insufficient detail, wrong codes or specifications, and errors in the drawings.

For your new home, take time to find a good architect/architectural draftsperson. The majority of draftspeople contribute their talents to solving today's rural and suburban residential problems: the problems the average family faces when they plan to build a new home. Architects on the other hand have the training to design commercial buildings, apartment high rises, etc. Because of their complex training and certification they are frequently not price competitive with a draftsperson. Sometimes they charge a percentage of total costs.

If you intend to use *stock plans* to construct your home, be sure they are the work of a reputable architect/draftsperson.

When you select an architect/architectural draftsperson, research references from other builders, and study the quality of the work compared to the costs. Ask about construction experience.

A good architect/architectural draftsperson should:

(a) Provide a free *estimate* regarding the cost of your plan
(b) Have experience in framing and construction
(c) Provide insight into interior floor plans and give recommendations for changes
(d) Understand the costs associated with alternative construction methods, materials, and floor plan (e.g., the economy of placing one bath over or adjacent to another)
(e) Know or have access to the required codes and specifications for working drawings
(f) Know what details are required by mortgage lenders and building permit departments pertaining to drawings
(g) Keep work neat and easy to read by others (trades, suppliers, etc.)
(h) Provide additional detail where required - will not just assume construction details to be understood by trades (e.g., beam construction, truss detail, floor joist layout, etc.) You should discuss options with your architect/architectural draftsperson.
(i) Draw a plot plan
(j) Double check all measurements to guarantee no costly mistakes during construction

When you talk to an architect/architectural draftsperson about designing your house, take the following material with you.

(a) A copy of the map showing your lot and legal address - for preparation of a plot plan, also bring the development plans if available
(b) A proposed sketch of the interior floor plan and exterior front elevation
(c) A list of questions to assist you in discussing your plan thoroughly
(d) A list of products and construction methods you intend to use for construction and finishing (material specifications). For example, asphalt shingles, 2 x 6 walls, fibreglass batt insulation, forced air furnace, 2 x 10 floor joists, etc.
(e) A picture of the property showing the slope of the site and adjacent properties. This may be helpful for planning the location of basement windows
(f) Before beginning preliminary sketches, obtain all rules and regulations governing the building activity in your area (local building codes, local zoning restrictions, local city sanitary requirements, developers' architectural controls, etc.)

So, how much should I spend on plans?

PLANNING TO MAKE YOUR HOME PASSIVE SOLAR EFFICIENT

You've got to watch out for yourself! There are many frauds in the business of hustling energy products. You have to decide which products you will recover the costs from and eventually receive paybacks from energy savings. The recommendations on the following page for saving energy are based on sealing the home tight and achieving passive solar gains.

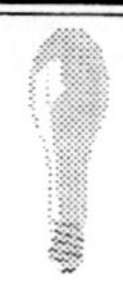

Some Ideas to Make Your Home More Energy Efficient

1) Preserved wood foundation
2) Strap exterior walls with 1 X 4, and use rigid insulation instead of 3/8 plywood sheathing on the exterior
3) Fibreglass batt insulation between exterior trusses around perimeter
4) R20 batt insulation in exterior walls
5) Use rigid insulation to prevent condensation from poly near exterior walls around subfloor
6) R40 - R50 loosefill insulation
7) Install a 6 mil polyethylene vapour barrier during the framing stage on ceiling and around subfloor
8) Acoustical sealant used to join poly
9) Poly, seal and staple around windows before installing
10) Strapping on inside of exterior walls to provide poly protection, supports for electrical boxes and 1 inch air space adds to insulation value. (use 2 x 2 and 2 x 4)
11) Strapping on ceiling with 2 x 4
12) Additional top plate on all exterior and interior walls to allow height difference for ceiling strapping
13) Shallow electrical boxes attached to strapping (weave wire behind strapping, no damage to poly)
14) High heel trusses used to provide room for additional insulation over exterior walls
15) Place weatherproof cardboard insulation stops between trusses to prevent insulation from entering eaves
16) Attic hatch on outside gable end for air tightness and to prevent loss of heat
17) Air to air heat exchanger
    - preheats fresh air from outside
    - ventilates inside air
    - exhausts stale air
    - keeps heat in
18) All bathroom fans and vents duct back to air exchanger, charcoal filter type range fan
19) Smaller, more efficient furnace (approx. 50,000 BTU)
20) Airtight wood stove used as an alternative heat source
21) Double glazed windows, casement or sealed units on south side; minimize windows on north elevation
22) South exposure - protect north elevation with garage, trees
23) Double front entry system

Other

- Continuous vapour barrier provides air tight home
- Well caulked and sealed
- Roof overhang to limit summer sunshine into house
- Thermostat controlled fan to circulate warm air and prevent overheating

EVALUATE YOUR PLAN

The following questions may assist you in evaluating your proposed house plan.

- Does the plan provide shelter over the front entrance?
- With doors open, is there room in the entrance area for two or more people?
- Is there a coat closet in the front entrance area?
- Will your furniture fit, leaving enough room for family activities?
- Is there a fireplace and can you group furniture around it?
- Are electrical outlets conveniently located for radio and TV, reading lights, cleaning, and other equipment?
- Will heat registers and cold air returns affect furniture placement?
- Is the dining area convenient for serving meals?
- Is there room for daily needs and entertaining?
- Is there convenient access to the patio or back yard?
- Is there sufficient floor space to accommodate a buffet?
- Mentally prepare a meal. Will the kitchen be convenient to work in?
- Is there adequate space for two people to work?
- Is there adequate storage space?
- Is counter space large enough and convenient?
- Is there a place to keep and operate appliances?
- Has provision been made for a dishwasher and microwave?
- Will your present equipment fit into the kitchen?
- Will the lighting be bright enough? Is there a separate light over the sink area?
- Are electrical outlets handy? Are there enough of them?
- Is there easy access to the dining area and front or back doors from the kitchen?
- Is there space to serve and enjoy meals in the kitchen?
- Are all hallways wide enough and will they be well lit?
- Is there wasted hallway space?
- Is there adequate storage space for linens?
- Are the stairs too steep?
- Will there be enough headroom coming downstairs?
- Is an electrical outlet placed to allow you to use a vacuum on the stairs?
- Is the capacity of the furnace and water heater adequate?
- Are there enough bedrooms and are they of sufficient size for your furniture?
- Does the plan meet your objectives and needs?
- Will the design be readily accepted in the marketplace?
- Is the plan affordable?

ELEMENTS OF KITCHEN DESIGN

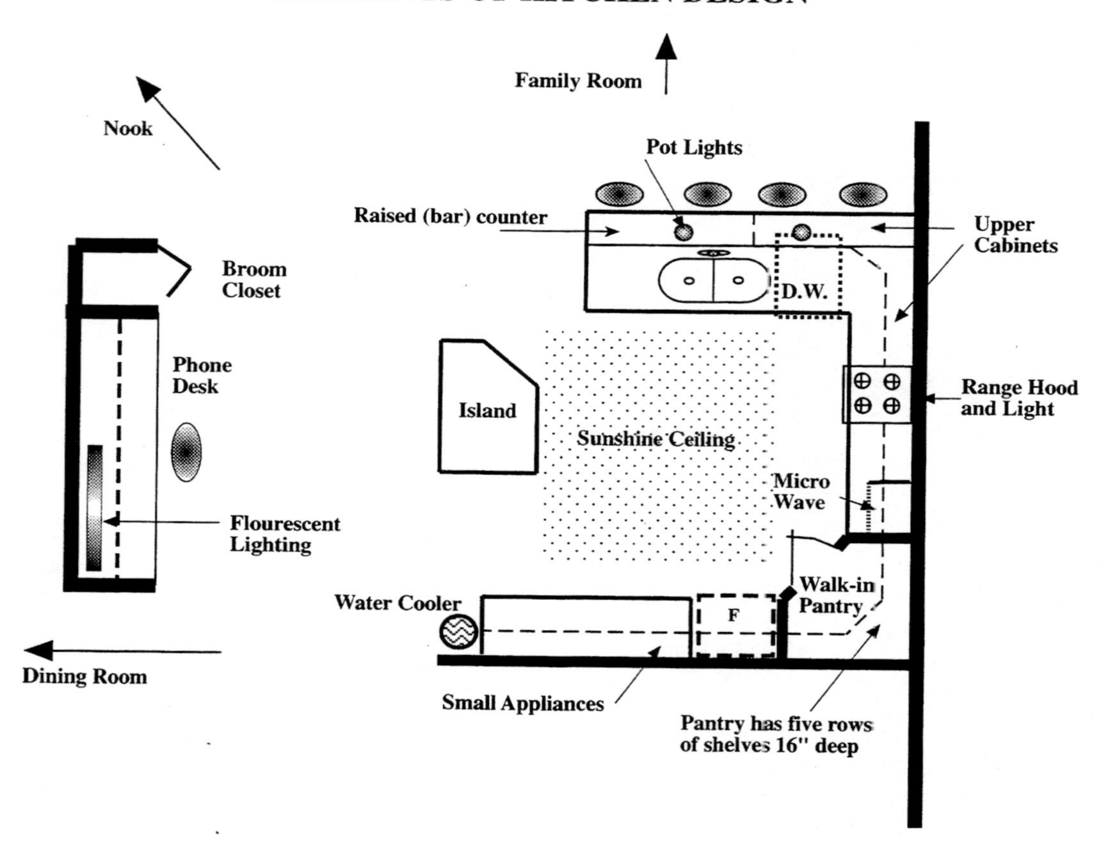

There can be no arguments to the fact that the kitchen is a very important part of your home, especially if you have a large family. I'm sure you can remember spending the whole day in the kitchen preparing meals and cleaning. Spend extra time carefully designing your kitchen to suit your lifestyle and needs. Here are some pointers to consider.

- Storage space - A walk-in pantry can give you 40 lineal feet of shelving space 16" wide. Place the pantry in the corner to provide practical use of dead space in the kitchen.
- Traffic flow - Design your kitchen for the easiest possible traffic flow. In our home the kitchen is in a central location with the family room, nook, and dining room fanning directly off the kitchen in three different directions.
- Location of appliances - Think of the easiest way to work in the kitchen. After dinner the dishes are placed by or in the sink, so you need some counter space here. Next they are rinsed off and placed in the dishwasher, so the dishwasher should be right next to the sink. After the dishes are cleaned in the dishwasher they are placed in a cabinet which is located directly above the dishwasher avoiding having to carry dishes around the kitchen. You want to be able to clean up quickly, right!
- Mentally prepare a meal to help you work out any bugs in your kitchen design.
- Play around with your design and consider alternative layouts on paper before you make your final selection.

How Important is the Front Elevation !

Many of the money saving ideas presented in chapter two delt with having a design that will give you a high appraised value. Curb value is the most important of all the ideas presented. The front elevation will reflect the value of your property therefore it will require your special attention to every detail. Here are some examples of front elevations showing different roof lines and details. The detailed elevations are the copyright property of Spectrum Design and Drafting Studios, Didsbury Alberta.

Two Story Estate Home

Two Story Country Home

Bungalow (Attractive entrance, cottage roof lines and gables)

Two story (den and living room face front)

Bungalow (Designed to look taller with higher roof)

Before

After

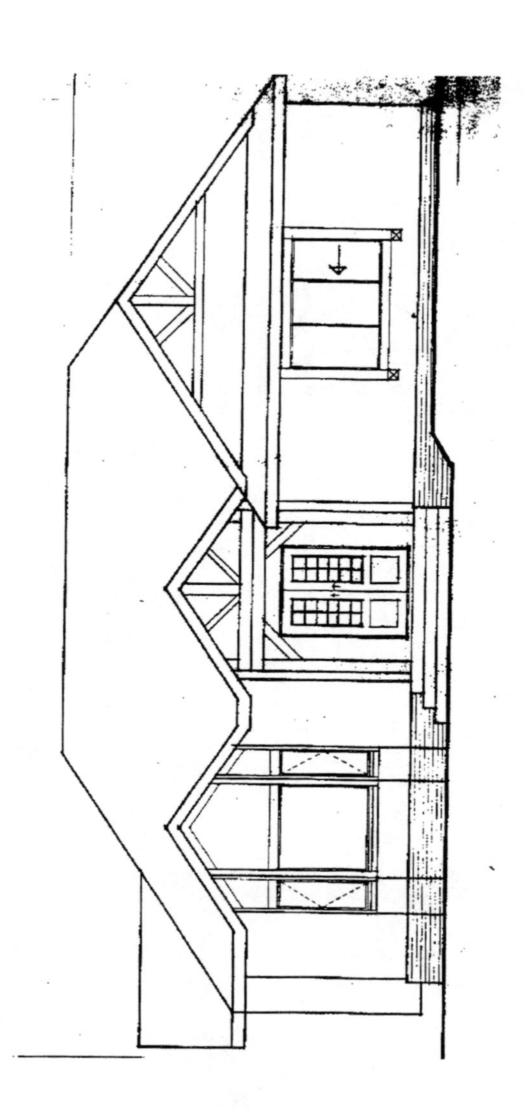

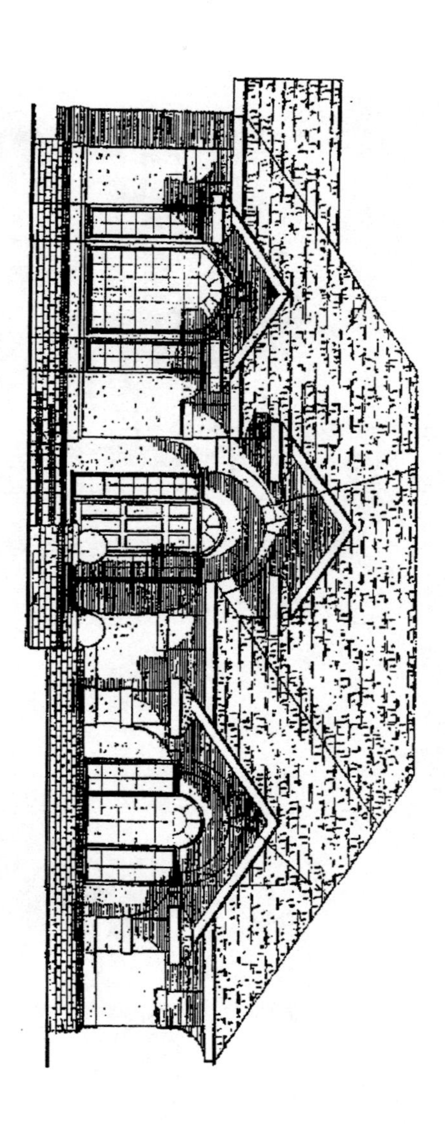

It is obvious, looking at these two different elevations, of the same home, that the appraised value or future resale value can be greatly affected by the appearance of the front. You can renovate the front of your home to improve appearance and increase value.

A

(roof slopes from front to back)

B

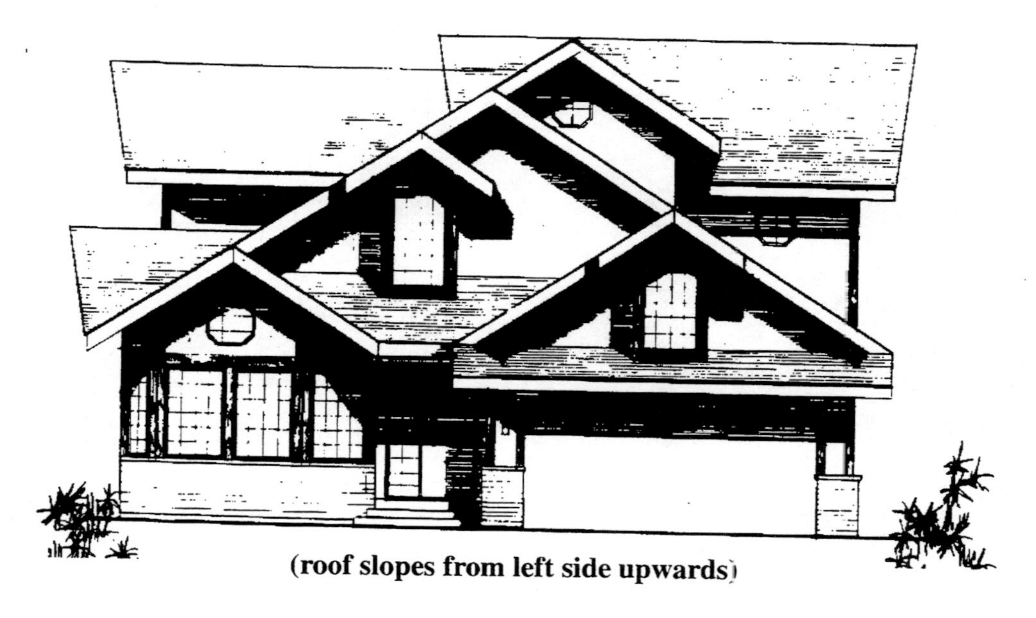

(roof slopes from left side upwards)

Both A and B have the exact same floorplan. The two completely different front elevations prove that you should concern yourself first with the floor plan. The front view, as illustrated here, can be altered to your liking afterwards.

APPLYING FOR A BUILDING PERMIT

You will be ready to apply for a building permit after purchasing the land and completing plans. The cost will vary depending on the amount of developed area. You can check the costs with the building permit office of your city planning department.

1. Plans and specifications for working drawings

Your plans must be clear and legible and contain all the specifications necessary for checking the foundation, framing, mechanical, venting and other details which may require city hall examination.

The following checklist can be used as a guideline to ensure your plans and specifications are complete.

(a) Plot plan

- What are the dimensions of the lot?
- What is the distance between the house and lot lines (side yards and setbacks)?
- The legal survey address of the property
- The street or civic address
- The scale used to draw the site plan
- Location of any easements
- Location of services (if required)

(b) Grade slip

- The lowest top of footing
- Suggested front grade
- Height of sanitary sewer invert
- Elevations at corner points of lot and house
- Arrow showing slope and direction of drainage

(c) Blueprints

General

- Name, type, and location of building
- Name of owner or builder
- Name of architect or designer and address
- Name of engineer (if required)
- The scale used for plans
- The date plans were drawn and revisions if any
- North point on plans
- Dimensions and height of all rooms
- Intended use of rooms

Footings and foundation

- Strength of concrete footings and walls
- Depth and width of foundation footing
- Size of rebar and location in footing and foundation walls (two rows generally top and bottom in walls)
- All footings to be below first line - minimum 4" or 1200 mm
- Depth and width of supports for the beam and other column pads
- Description of the support under the landing of the basement stairs
- Tile drains in accordance with required drainage and local water table
- Description of the foundation wall construction
- Concrete form tie holes individually sealed and two coats dampproofing on exterior of concrete walls
- Require vapour barrier under concrete floor slab
- What is the thickness of the basement floor slab?
- How thick is the concrete driveway and how will it be reinforced?
- The number and size of piles that are shown

Ventilation

- Location of dryer vent
- Location of fresh air intake and required distance from day vent
- Size, shape, and height of chimney or gas vent
- Size and location of combustion air and other ventilation openings
- Location of bathroom and kitchen exhaust fans
- Size and continuity of all pipes, ducts, shafts, flues, and fire dampers

Framing

- Description of the floor joist system including bridging
- What type of beam is specified for the basement?
- Are there any bearing walls on the main floor?
- Description of the construction of the exterior and interior walls and the ceiling including insulation
- Use of joist hangers around all floor openings prior to placing sub floors
- What is the size of the attic access?
- Detailed description of beams and headers
- Type of exterior sheathing
- Grading of all lumber which must be #2 or better
- Detail of windows and doors including size and weather stripping (must be government approved)

Mechanical

- Where the furnace is to be located and what size it is
- How many and location(s) of outdoor waterproof electrical outlets
- Electrical layout showing all wall outlets, switches, and lighting fixtures
- Location of electrical panel
- Location of washer stand pipe
- Location of smoke detectors
- Location of interior and exterior water service (basement)
- Location, size, capacity, and type of all principal units of equipment
- All mechanical to meet municipal by-laws and provincial codes

Roofing
- Pitch of the roof
- Type of roofing specified
- Detail showing construction of the members of trusses or rafters

Siding
- Describe the exterior finish of the house
- What type of soffits are specified
- All flashing, caulking, eavestroughs to meet requirements

Others
- All floor coverings to be CSA approved
- All appliances to be CSA approved
- All closet rods over four feet long to have centre support
- Stairs, handrails, guards to meet residential standards

2. What to bring

The following information is generally required to submit for a building permit.

(a) Two sets of building plans which must conform to the local codes. These plans are to be complete, legible, and drawn to scale. Check with the city or town office to see what scale they will accept.

Common scales: Imperial 1/4" = 1'Metric 1 : 50

Even though you require complete plans and specifications it is also important to keep your package as simple as possible. Do not provide them with information they don't require. I was recently told by a plan checker at city hall that a one-page set of plans can usually be processed faster than a ten-page set of the same plans.

(b) Two copies of a site plan (plot plan) of the property (see Chapter 7)

(c) A builder grade slip (see Chapter 7)

(d) One copy of the certificate of title available from the land titles office. It is your responsibility to ensure the work being carried out does not contravene the requirements of restrictive covenants, caveats, or any other restrictions registered against the property.

(e) Payment for the building permit.

In summary, getting your building plans through city hall should be looked on as a simple procedure. The plan checker, who may be the building inspector, generally has a good background in construction and knowledge of codes. The main purpose of checking your plans is to ensure that a fire safe and structurally safe building will be erected. In order to carry out this responsibility, the checker must thoroughly examine the plans, check bearing loads, etc., ask you questions, and maybe make some recommendations for changes which will likely be for the better. If your plans are drawn by an experienced draftsperson or an architect with construction knowledge, you will likely have few problems with city hall.

Additional Tips

(1) If your plan is complex, with many beams, you should consider having your beams sized and specified (i.e. 3 ply 2 x 12, microlam, steel) by an engineer. The engineer will stamp your plans with his professional engineer stamp enabling quicker and easier processing through city hall or township office.

(2) If you're building on an acreage site contact the township for building permit requirements. You may need additional plans for your well and septic system.

11

ESTIMATING

Estimating can be the most important step in successfully building a home. Imagine yourself as a building contractor. You must have work, and you must have it at a price that yields a profit, otherwise sooner or later you'll be out of business. To be successful, the builder must have an understanding of all costs so that bids are not too high or too low. When the price is too high, you will lose the job to a competitor and when it is too low you run the risk of being below your actual costs and losing money.

It is similar for the owner builder. When you underestimate the cost of your home, you will exceed your budget for mortgage purposes. If you spend too much by over estimating you will not save any money, hence no profit. To avoid this you must have a knowledge of all costs associated with your future home.

Every owner-builder will be involved in estimating to a certain degree. The amount of work will vary from just picking up estimate bids from contractors and suppliers to doing an actual material take-off of building materials. (A material take-off involves calculating the quantity of construction materials and quality to be purchased. It is compiled by taking information from the blueprints and applying it to a standard rule of thumb or formula.) The take-off should be done by an experienced estimator, estimating service company or the estimator from the lumber yard. Formulas are provided in this chapter for those with some construction knowledge who would like to do their own material quantity take-off.

The material quantity take-off will be your own specific materials list to deliver or fax to the lumber yards. The tenders you receive, provided you specified quantity and quality, will be very easy to compare. When you begin to price cribbing, framing and mechanical trades, all you need to do is submit plans and specifications for pricing.

Qualifications of a Good Estimator

To do a good job of estimating you should have:

(a) the ability to read plans and measure if you are estimating quantities.
(b) some general knowledge of construction materials and methods.
(c) the ability to see the project being built in your mind, and
(d) an average amount of common sense (be conscientious, ask questions and check all figures)

If you don't feel confident that you possess enough construction knowledge to do a fairly accurate job of the material take-off, seek help from those in the industry who have experience. The best person to help you would likely be the Estimator at your local lumber yard. Otherwise an estimating service company such as Central Estimating Service in Calgary, Alberta can be used for doing your material take-offs (lumber only). Having an Estimator provide this service saves you time which you can use for researching all other costs to quickly price out the project.

To complete your own quantity take-off, you will need a good set of working drawings, a columnar pad and pencil, ruler or architect scales (metric and Imperial) and a standard calculator. You also need to know these basic definitions.

ANGLE IRON: An L-shaped steel bar frequently used to support masonry.

BEAM: A means of supporting a floor or roof system, e.g., 3 or 4 ply 2 x 10 planks nailed together would form one type of beam.

BEARING WALL: A wall that supports any vertical load in addition to it's own weight.

BLOCKING: Small pieces of framing between joists or studs for support. Lumber scraps 16" or longer should be saved and used for blocking.

BRIDGING: Cross bridging or blocking used to prevent floor joists from warping sideways.

BUCK: A wood frame placed in a cribbing form to create an opening in the concrete for windows.

CANTILEVER: A construction unsupported at one end that projects outward to carry the weight of a structure above, such as a balcony, bow window, extension for china cabinet or fireplace chase.

CASING: A form of trim used around windows and door openings.

CAST-IN-PLACE: When constructing a foundation, cast-in place means embedding the perimeter of the floor system into a concrete foundation wall. Often the subfloor rests on top of the foundation which is not cast-in place.

CRIPPLE STUDS: Short framing members from the bottom plate to sill, examples are the short wall studs below windows.

DOUBLE JOISTS: Two joists together in a floor system (required for extra support or under interior partitions running parallel to joist system)

FASCIA: The finished edge of a roof around the face of eaves and roof projections

FLOOR TRUSSES: An alternative to floor joists in a subfloor system which allows passage of ductwork with a clear span in basement

FOOTINGS: The base of a foundation wall or pier, usually made of concrete

FORM TIES: Thin metal strips used to prevent the spread of form sides while pouring and curing a concrete basement

FROST WALL: A foundation wall designed to prevent the passage of frost; it is usually built deeper into the ground on a lower elevation of the foundation

GABLE: The upper triangular-shaped portion of the end wall of a building

GALVANIZED IRON: Dipping iron into molten zinc to protect it against rust, used for flashing, nails, and other materials requiring protection

INSULATION STOP: A waterproof piece of cardboard placed between roof trusses to prevent blown-in insulation from entering the eaves

JAMBS: The side member of lining of a doorway, window or other opening

JOISTS: A horizontal plank supporting a subfloor or deck (common 2 x 10 or 2 x 12)

JOIST HANGERS: A metal bracket used for additional support at the intersection of floor joists, may be single, double or triple

LINEAL FEET: A term used in measuring an amount of lumber (LFT)

LINTEL: A load-bearing horizontal member directly above the rough opening of a door or window (common 2 or 3 ply 2 x 10)

MUDSILL: The lowest horizontal plank, which serves as a base for other framing members

PARGING: A form of external plastering, a mortar mixture, often applied on the foundation walls below the siding or exterior base

PERIMETER: The distance around the outside walls

PIERS: A column of masonry used to support other structural members

RANDOM LENGTH: A term used to describe ordering a quantity of lumber when exact size of each piece is of no concern (R/L)

REBAR: Reinforcing steel rods used to strengthen a foundation

RISE: Refers to vertical distances in stairs (i.e., number of risers) and used for measuring slope in a roof (rise is the distance straight up from top of wall to peak)

SHEATHING: Wall covering applied directly to the exterior of the frame

SILL: The lowest horizontal member forming the bottom of the rough opening of a window or door

SLOPE: The measurement of the vertical rise to the horizontal run, e.g., 3 in 12 slope

SOFFITS: The finished surface of the underside of a roof overhang or eaves

STRINGERS: The supporting sides for a set of stairs usually built from 2 x 10 or 2 x 12 material

STRONGBACK: In framing a corner, a strongback is a vertical stud placed at a 90 degree angle to give added strength

STUD: A full-length upright member (common 2 x 4 or 2 x 6) extending from bottom plate to top plate

TELEPOST: A steel adjustable post used to support a beam

TOP AND BOTTOM PLATES: The horizontal members in a wall, which are nailed to the top and bottom of the wall studs

TREADS: The horizontal member of a set of stairs; the part you walk on

TRIMMER JOIST: A joist that trims the outer edge of a subfloor (common 2 x 10 or 2 x 12)

TRUSSES: A structural framework of members arranged and fastened together to span and support a roof

UNDERLAY: An additional layer of sheathing placed under lino or tile

VAPOUR BARRIER: A polyethylene material used to prevent the passage of air, vapour or moisture

WEEPING TILE: Drain tiles or perforated plastic pipe placed around the footings to collect and drain away water

FIVE STEPS TO ESTIMATING

Start your estimating ***first*** by studying the plan and the specifications on the plans. This will give you a good idea of the materials the house is constructed from and will enable you to recall details when working through your checklist or doing a material take-off.

The ***second*** step is to make a list of the type, quality, and grades of materials to be used. Use the sample specifications list provided to get started (see **Worksheet #17**). You will want to specify this information when seeking tenders for materials, for example, 2 x 6 kiln-dried construction grade spruce studs, 2 x 10 No. 1 douglas fir, tongue and groove 5/8 fir plywood. Some of the other items you will need to select are, the type of heating system, fireplace, all interior finishing materials, exterior finish and roofing materials.

The ***third*** step is to obtain a materials list for all the lumber. Only those people with construction experience would take on the challenge of preparing a material quantity take-off to determine the estimated amount of construction materials (see **Worksheet #18**). The formulas provided in this chapter would only be used for an addition, a renovation, for reference to do a comparison or for a person with framing experience.

For a complete home, the material quantity take-off should be done by an experienced estimator. Most lumber retailers employ estimators trained to provide this service for a small fee. I highly recommend you develope a good working rapport with your lumber supplier because you will be counting on his (her) assistance for the timely delivery of the various packages of lumber to your site.

Make an appointment to review your plans and discuss your needs. You will need to set up a line of credit with your major lumber supplier so that you do not have to pre-pay for materials or be on site to pay for them when they are delivered. They will require proof that either your mortgage is approved or you have alternate interim financing or other money available to pay the invoices you will be receiving.

The ***fourth*** step is to complete a detailed estimating checklist (**see Worksheet #20**). You will need to obtain tenders on all major items and judge the remainder for yourself. This will require a great deal of work, but can lead to substantial savings and peace of mind knowing an approximate final cost before you begin. Try to obtain actual estimates, or tenders, for items 2 to 17 and item 22.

The ***fifth*** and final step is to summarize the estimating checklist on a simpler form for your mortgage application. The **Cost Summary and Loan Calculation** form summarizes all your estimates to give you a grand total cost for your new home or renovation. This important form is located in the estimating checklist section under **Worksheet #20 page 11-25.**

Worksheet # 17

SPECIFICATIONS

NAME: ______________________________

BUILDING ADDRESS ______________________________

1. **FOOTINGS:**
_ Walls 18" x 6" concrete
_ Telepost pads 32" x 32" concrete to support beam
_ Concrete curbs flush with top of basement floor for bearing walls
_ Concrete piles to support deck
_ Concrete piles under garage floor
_ Frost protection footings for walk out basement
_ All footings reinforced with rebar if required

2. **BASEMENT WALLS:**
_ 8' X 8' concrete
_ Two rows rebar top and bottom of wall
_ 2 x 4 ladder on top of wall
_ 2 x 6 ladder on top of wall
_ 2 x 8" cap plate on top of concrete wall
_ Joists cast into concrete wall (cast-in-place)
_ Snap tie holes filled and two coats foundation tar-to-grade
_ Exposed concrete walls parged
_ All concrete to meet standards with necessary additives

3. **BASEMENT FLOOR:**
_ 4" concrete finished smooth
_ Placed on 6" compacted gravel
_ Poly laid on top of gravel as moister barrier

4. **GARAGE FLOOR:**
_ 4" concrete finished smooth
_ Reinforced with rebar or wire mesh
_ Placed on 5" compacted gravel

5. **DRIVEWAY:**
_ 4" concrete brush finished with control joints
_ Reinforced with wire mesh
_ Placed on 5" compacted gravel
_ Poured as wide as garage

6. **FRONT WALK:**
_ Poured concrete from driveway to step
_ Control joints where necessary
_ Reinforced with rebar

7. **FRONT STEP:**
_ Precast front step bolted to house
_ Poured concrete step, brushed finish and reinforced

8. **OTHER CONCRETE:**
_ Concrete block walkway to house
_ Concrete planter base
_ Concrete grade beam
_ Retaining wall
_ Concrete deck brackets, front step and walkway supports
_ Concrete pad for back garage door
_ Concrete pad for walk-out basement
_ Concrete steps and wall for walk-out entry
_ Concrete brick ledge
_ Concrete base to support outdoor heating system

9. **FRAMING MAIN FLOOR:**
_ Beams 2 x 10 construction grade Douglas Fir
_ Laminated plywood beams (microlam) where required
_ 3" steel teleposts where required
_ Joists 2 x 10 construction grade #1, Douglas Fir or 9 1/2" floor trusses
_ Flooring 5/8" tongue and grove Fir plywood or OSB board (Oriented Strand Board)
_ Subfloor heavily nailed and glued (Screwed, optional)
_ Exterior walls 2 x 6 construction Spruce 16" on center
_ Interior walls 2 x 4 construction Spruce 16" on center
_ Bearing walls 2 x 6 construction Spruce 16" on center
_ Exterior sheathing 3/8" Spruce Plywood or OSB (Oriented Strand Board)
_ All window headers 2 x 10 Douglas Fir
_ All crossbridging and blocking

10. **FRAMING SECOND FLOOR:**
_ Joists 9 1/2" or 11 7/8" Floor trusses placed 24" on center
_ Microlam beams where required
_ Flooring 3/4" tongue and groove Fir plywood or OSB board
_ Joists 2 x 10 construction grade No.1 Douglas Fir
_ Flooring 5/8" tongue and groove Fir plywood or OSB board
_ All walls framed same as main floor

11. **ROOF FRAMING:**
_ Engineered approved trusses spaced 24" on center
_ Tile roof 2 x 6 top cord on trusses, 3/8" sheathing, strapping
_ Shake roof - 1/2" sheathing OSB Board or plywood
_ Asphalt shingle roof 7/16" plywood sheathing

12. **STAIRS:**
_ Built by your the carpenter or
_ Pre-manufactured stairs - made to site measure
_ 1" plywood treads
_ Routered, nailed and glued

13. **DECK FRAMING:**
_ Joists 2 x 10 construction Fir
_ Flooring 2 x 4 construction Spruce
_ Flooring 2 x 4 Cedar
_ Flooring 3/4" Spruce ply coated
_ Railing 2 x 4 and 2 x 6 Spruce
_ Railings 2 x 4 and 2 x 6 Cedar
_ Pickets 2 x 2 construction Spruce
_ Pickets Cedar
_ Railing, built-in bench and backrest of 2 x 4 construction Spruce with 2 x 6 top rail

14. **OTHER FRAMING:**
_ Basement walls studded - 16" on center
_ Scissor truss vaulted ceiling
_ TJI full vaulted ceiling
_ Sunken Living Room
_ Raised Dining Room
_ Fireplace chase
_ Bay Window
_ Box window
_ Drop framed kitchen ceiling
_ False beams
_ Strapping for vaulted ceiling
_ Pocket door framing
_ Framing step and around Jacuzzi

15. **ROOFING:**
_ Interlocking asphalt shingle roof
_ Heavy certified #1 shake roof
_ Estate tile roof
_ Economy tile roof
_ All flashing, caulking, vents and poly

16. **EXTERIOR DOORS AND WINDOWS:**
_ Wood double glazed casement with crankout handles
_ Double glazed awning windows with lever crank handles
_ Double glazed wood sliders
_ Double glazed sealed units
_ Wood window grills on front of house
_ Front door - Steel insulated
_ Front door - Oak
_ Patio Doors
_ French style entrance doors
_ Attic vents
_ Garage door to house steel slab
_ Front garage door raised panel stain grade
_ Front garage door paint grade Mahogany
_ High security brass plate and ball handle front door lock
_ Brass lever front door lock
_ Optional exterior vinyl cladding

17. **EXTERIOR FINISH:**
_ Stucco Finish
_ Pre-stained cedar siding
_ Double 4 Aluminum siding
_ Cedar corner battons and trim
_ Fir trim
_ Caulking around all joints
_ Fascia Aluminum
_ Fascia Cedar pre-stained
_ Soffit Aluminum venting
_ Soffit channel Cedar pre-stained
_ Soffit 1/2" Fir plywood
_ 5" Aluminum eaves and downspout
_ 6" Aluminum eaves and downspout

18. **EXTERIOR TRIM**
_ Cedar battons as per plan
_ Brick trim as per plan
_ Stucco relief (trim) as per plan

19. **INSULATION AND DRYWALL:**
_ Exterior walls R20 Fiberglas batt insulation fit for 16" on center studding
_ Basement walls R12 batt insulation fit for 24" on center framing
_ Attic R40 Blown in insulation, Rockwood
_ 1/2" drywall on all interior walls and ceilings except basement
_ All ceilings to be stippled except bathrooms
_ Interiors of showers and around baths to be finished with water resistant drywall
_ Garage walls common with house to be 5/8" fireguard drywall
_ Other garage walls and ceilings to be 1/2" drywall
_ All drywall screwed in place
_ Drywall touch-up after finishing

21. **HEATING AND VENTILATION:**

ALTERNATIVE #1

_ Mid efficiency forced air natural gas furnace(s)
_ All grills and registers
_ Outdoor venting of hot water tank(s), furnace, dryer and bathrooms - range hood optional
_ Power humidifier for furnace(s)
_ Standard thermostat

ALTERNATIVE #2

_ Airco high efficiency gas furnace(s)
_ All grills and registers
_ Outdoor venting of hot water tank(s), furnace, dryer and bathrooms - range hood optional
_ Power humidifier for furnace(s)
_ Electronic programmable thermostat

ALTERNATIVE #3

_ High efficiency (ICG Ultimate or Lennox Pulse) flueless forced air gas furnace (no pilot light)
_ All grills and registers
_ Outdoor venting of hot water tanks(s), furnace, dryer and bathrooms - range hood optional
_ Power humidifier for furnace
_ Electronic programmable thermostat

ALTERNATIVE #4

_ Amana high efficiency heating system, flueless and no pilot lights
_ Amana high efficiency water heating system, flueless and no pilot lights
_ All grills and registers
_ Outdoor venting of furnace, dryer, bathrooms - range hood optional
_ Power humidifier for furnace(s)
_ Electronic programmable thermostat
_ NOTE: No standard hot water tanks to be supplied by plumber

ALTERNATIVE #5

_ Hot water heating
_ Underslab heating, Radiant Floor heating (boiler)

20. **PLUMBING:**

MAIN BATH UPSTAIRS

Colored fixtures selected by owner

_ American Standard (A.S.) 5' Salem bathtub
_ Delta Single Lever tub filler and shower faucet
_ A.S. Cadett #2 rim toilet and seat
_ Delta single lever washerless 4" center set taps with pop up drain
_ A.S. Ovation sink (Marble-optional) - color by owner

ENSUITE BATH

Colored fixtures selected by owner

_ 2" shower drain with rubber liner
_ Delta single lever taps, A.S. Ovation sink
_ A.S. Cadett #2 toilet
_ A.S. 6' bathtub

MAIN FLOOR BATH

Colored fixtures selected by owner

_ A.S. Ovation sink
_ A.S. Cadett #2 rim toilet and seat
_ Delta single lever 4" center set taps and pop-up drain

KITCHEN

_ Swanstone sink
_ A.S. double stainless steel kitchen sink
_ Delta single lever deck faucet with spray
_ Delta (white) decorative deck faucet
_ I.S.E. # _____ garburator
_ Connection of dishwasher

LAUNDRY ROOM

_ 2" Washer drain
_ Hot and Cold washer taps
_ Stainless steel sink with bar faucet
_ Single laundry tub with legs
_ Emco laundry tub facet

BASEMENT

_ 4" Floor drain
_ 3 piece rough-in for future development
_ Gas line to furnaces
_ Two 1/2" non-freeze lawn services
_ 3/4" Copper water services
_ All necessary valves and piping, drain and vent pipes and cleanouts
_ Two 33 gallon (40 US Gallon) Hot water tanks Model # 402
_ One 50 gallon Hot water tank
_ Water softener hook-ups

22. **ELECTRICAL:**

(Temporary power during construction)

_ Underground electrical service to house
_ Circuit box as required
_ Copper wire throughout
_ Smoke detectors installed and wired to circuit box
_ 220 volt outlets for stove and dryer on main floor
_ 4 weather proof outdoor outlets as per drawings
_ Ceiling plug-in in garage for door opener
_ Wiring for dishwasher and garburator
_ Ground fault razor outlets in Master Bedroom and Main Bathroom
_ Exterior exhaust fans in all bathrooms - except bathrooms with windows or as specified by owner
_ 4 telephone jacks (location by owner)
_ 4 cable outlets (location by owner)
_ Outlet and switch plates in White (Ivory)
_ Hanging of owners fixtures
_ Generally, all electrical as indicated

OTHER

_ Basement rough in
_ Recessed lighting in kitchen
_ Lighting in steps

23. **FIREPLACES:**

ALTERNATIVE #1

_ Gas fireplace
_ Only small fireplace chase required
_ Electrical fan hook-up, gas hook-up
_ Finished in Oak, Granite, Marble, etc.

ALTERNATIVE #2

_ CSA approved metal box fireplace with fresh air intake
_ Fireplace facing in brick or stone - floor to ceiling
_ Raised hearth
_ Oak mantel
_ Exterior of chase finished same as outside exterior

ALTERNATIVE #3

_ Energy efficient airtight heat circulating fireplace Glowboy or Pressurizer with fresh air intake
_ Fireplace facing in brick or stone - floor to ceiling
_ Raised hearth
_ Oak mantel
_ Exterior of chase finished same as outside exterior

ALTERNATIVE #4
OPTIONAL ITEMS
_ Masonry fireplace
_ Masonry chase
_ Wood stove
_ Formal finished fireplace in Oak or Ash
_ Combination of any of the above

24. **INTERIOR PAINT:**
_ One coat of primer sealer 1/2 tinted to final paint colour
_ All areas two coats of Latex paint on walls
_ Interior of all closets and shelves painted two coats Latex
_ NOTE: Finish stain and lacquering under No. 27 INTERIOR FINISHING

25. **EXTERIOR PAINT:**
_ Window trim, one coat primer followed by 2 coats quality stain
_ Battons 2 coats stain
_ Garage doors pre-stained one coat

26. **KITCHEN CABINETS:**
_ Kitchen cabinets, bath vanities, laundry room cabinets, counter tops and all accessories
_ supplied and installed by manufacturer or seller

27. **INTERIOR FINISHING:**
ALTERNATIVE #1 - MAHOGANY
_ 2 1/4" Mahogany casings around all doors and windows
_ 2 1/4" Mahogany baseboards
_ Mahogany doors and bi-folds
_ Finished with matching stain, filled, sanded and two coats of lacquer
_ Mahogany railings and handrails
_ Exact stain color by owner
_ Toning on wood optional

ALTERNATIVE #2 - COLONIAL STYLE
_ All casings around doors and windows finished smooth to a crisp white finish with two coats lacquer
_ Baseboards white lacquer or Oak stain
_ White lacquer inset panel doors (colonial style)
_ Railings stained and finished in Oak

ALTERNATIVE #3 - OAK
_ 2 1/4" Oak casings around doors, windows and all baseboards
_ Oak slab doors and bifolds
_ Finished with matching stain, filled, sanded and two coats of lacquer
_ Oak railings
_ Exact stain color by owner
_ Toning on wood optional

ALTERNATIVE #4
_ Combination of above or alternative style casings and baseboards ie: white colonial upstairs, oak in family room.
_ Alternate mirror closet doors
_ Alternate solid Fir doors with Fir or Oak casings, or solid Oak doors

28. **HARDWARE:**
_ Brass ball shaped door handles on all interior doors
_ Brass pull handles on all bifolds
_ Dead bolts on front doors, other exterior doors mastered to the same key
_ Large main entrance brass door handles
_ Brass keyed locking door handles on steel garage doors
_ Brass door stops

29. **CERAMIC TILES:**
All styles selected by owner
_ Ceramic tiles 5' above tub in main bath (tiles to ceiling optional)
_ Ceramic around jacuzzi, 2 rows
_ Ceramic tile shower
_ Ceramic tile foyer 12" x 12"
_ Ceramic tile kitchen backsplash from countertop to bottom of upper kitchen cabinets

30. **BATHROOM ACCESSORIES:**
_ Full corner length 36" mirrors above bathroom vanities (exact size to be measured)
_ Chrome or brass colored towel racks and toilet paper holders in all bathrooms
_ Medicine cabinets optional (exact style by owner) (check correct size)
_ Shower rod

31. **FLOOR COVERINGS:**
ALTERNATIVE #1
_ 36 oz 100% nylon carpet throughout except kitchen, bathrooms, foyer and laundry room

_ Foam underlay under all carpeted areas
_ Armstrong No-Wax Linoleum for kitchen and bathrooms

ALTERNATIVE #2
_ 42 oz 100% nylon carpet throughout except bathrooms, foyer and laundry room
_ 3/8" rubber under all carpeted areas
_ High grade No-wax linoleum for kitchen and bathrooms

ALTERNATIVE #3
_ 50 oz 100% nylon carpet throughout
_ 1/2" rubber underlay under all carpeted areas
_ Congoleum style floor coverings for kitchen, nook, bathrooms and laundry room
_ Manington Gold Linoleum

ALTERNATIVE #4
_ Oak flooring
_ Combination of above, alternate weight or make of carpet

32. **EXTRA ITEMS:**
_ Gas loglighter, gas fireplace
_ Rough-in vacuflow
_ Garburator
_ Light fixture allowance
_ Appliance allowance
_ Cedar on ceilings
_ Feature walls
_ Bookshelves
_ Closet shelving
_ Electric garage door opener(s)
_ Smoke glass shower door
_ Curved staircase
_ Whirlpool tub

_ High grade counter tops - granite or corian
_ All Vinyl windows
_ Metal or vinyl clad windows
_ Large deck
_ Basement framing
_ Basement electrical and heating
-- Home theatre

33. **OTHER:**
_ Backfill to grade and finish with top soil
_ All necessary permits and surveys
_ Insurance until occupancy or completion
_ Utilities to possession date
_ Measures for winter construction
_ Subtrades warranties and builders warranties
passed on to owner
_ Legal fees to prepare construction contract but
excluding registration for mortgage purposes
_ All work to be performed by qualified and
experienced tradesmen
(preferably with Journeyman Status and
references)

WORKSHEET # 18

MATERIAL QUANTITY TAKE-OFF GUIDE

(For major materials only)

1. Footings

-Formwork

(2 x 6 or 2 x 8 material can be reused later in areas of framing)

Length of footing x 2 = lineal footage for both sides of footings.

Calculate telepost pads and bearing walls.

____

-Concrete

Volume = Length x Width x Height

27 cu. ft. = 1 cu. yard (convert all measurements to feet)

V. of cu. yds. = $\frac{L \times W \times H}{27}$ **add 5%**

____

-Rebar (two rows up and two rows down plus extra for corners)

Rows required x perimeter of footing = lineal footage of rebar

____

2. Foundation wall

-Concrete (convert all measurements to feet)

Volume = L x W x H

Round up to highest cu. yd.

(if metric measurement is required, refer to conversion tables in text)

____

3. Gravel (for under basement concrete floor)

-Find ground area plus 2 feet or .5m on all sides of footings.

-Multiply by the thickness of gravel required.

Volume of gravel = $\frac{\text{Area x Depth}}{27}$ **= cu. yds.**

Metric: Area x Thickness = cu. metre

____

4. Weeping tile

Perimeter of outside footing + 10 feet = lineal footage

-Draw out placement around house - you may require additional amount equal to the width of your house if weeping tile is placed through the centre. Check requirements!

____

5. Waterproofing (tar coating)

Area = perimeter of exterior x height of wall and footing

-waterproofing only goes to finished grade level.

6. Floor joists, headers, solid blocking, beam, lintels

-Floor Joists
Length of wall x 3/4 plus (assuming 16" on centre)
1 extra = joists required

-Lintels
Rough opening of each window or door opening plus 3" x 2 = length of material for each opening
Do this calculation for each opening.
If 3 laminations, then multiply by 3

-Beam
Have your draftsman, truss manufacturer or hire an experienced professional engineer to obtain proper lengths and specifications of materials to order.
These experienced people should be able to show you some different options of materials to use for beam construction i.e. wood, microlam, steel.

---

7. Cross Bridging

Number of spaces x 2
Note: -installed every 6 ft.
-not required for floor trusses
-ordered in bundles of 50

---

8. Subfloor plywood or oriented strand board (O.S.B. board) 4 x 8

Covered area (sq. footage of house) = sheets required
32 (sq. footage of 1 sheet)
Subtract for opening in stairway if significant size, e.g., some two story homes have large stairwell and foyer openings.

---

9. Wall framing (including blocking, backing)

-Exterior (2 x 6, 16" on centre)
Studs - one stud for every foot of wall length
Plates - length of wall x 3 = lineal footage for double top plates and single bottom plates.

-Interior (2 x 4, 16" on centre)
Studs - one stud for every foot
Plates - Length of wall x 3 = lineal footage
Note: Add extra 10% for drywall backing, blocking, bracing, etc.

10. Wall sheathing (4' x 8' plywood or O.S.B. board)

Number of sheets = total length of wall x height - area of major openings / 32

Note: Add extra for gable ends.

---

11. Roof sheathing

Number of sheets = area of roof / area per sheet (measure roof lengths on cross sections)

Or find ground area of roof which is area of house plus all overhangs (usually 2 feet on each side)

Convert ground area to roof area by adding percentage determined by slope

3/12	add	3% of area
4/12	add	6% of area
5/12	add	9% of area
6/12	add	12% of area
8/12	add	20% of area

---

12. Shingles

Total roof area / 32 = no. of bundles required

or one bundle for every 4 x 8 roof sheet

Note: you will require an additional 3 to 6 bundles, depending on the size of house, to be used for starter shingles.

---

13. Insulation

-Ceiling

Floor area = Sq. footage of ceiling insulation required.

-Walls

Area of walls less major openings = sq. footage required for walls.

---

14. Poly **Area of wall and ceiling + 500 sq. ft.** (round up to nearest roll)

15.Drywall **Floor area x 4 = approximate sq. footage of drywall**

Note: Actual board count and sizes should be done by the drywall installers or your drywall contractor.

---

16. Carpet

$\frac{\text{Floor area}}{9}$ + 5% = approximate number of sq. yds.

---

17. Miscellaneous

As per plans or check items on the material checklist or on the specifications list which are both provided in this book.

Worksheet # 19

LENGTHS OF DRYWALL SHEETS

(Use for ordering drywall)

Rooms	8 ft.	10 ft.	12 ft.	14 ft.	Other
Master Bedroom					
Bedroom # 2					
Bedroom # 3					
Bathroom					
Bathroom					
Living Room					
Dining Room					
Kitchen					
Hallways					
Foyer					
Other					
Other					
TOTALS					

Worksheet # 20

ESTIMATING CHECKLIST

(To be summarized on the cost summary and loan calculation form)

ITEM CHECK	PREPARATION	
	1. Blueprints, survey, etc.	
_____	Insurance	$_____
_____	Architects, draftsperson and blueprint copies	_____
_____	Plot plan	_____
_____	Estimate fees for material quantity take-off	_____
_____	Survey stakeout	_____
_____	Survey certificate (or real property report)	_____
_____	Footing elevation check and reset survey pins	_____
_____	Engineer soil bearing test (if required)	_____
_____	Engineer soil type test (if required)	_____
_____	Finished grade check (if required)	_____
_____	Other	_____
	SUBTOTAL	**$**_____

FOUNDATION

	2. Excavation, backfill, trenching, grading	
_____	Remove topsoil (if required)	$_____
_____	Excavation	_____
_____	Trenching for water, sewer and electrical	_____
_____	Move in equipment	_____
_____	Haul away excess dirt	_____
_____	Backfill	_____
_____	Bring site to rough grade	_____
_____	Bring site to finish grade	_____
_____	Deliver and spread topsoil	_____
_____	Landscaping (materials, decorative rock, etc.)	_____
_____	Other	_____
	SUBTOTAL	**$**_____
	3. Basement concrete, weeping tile, dampproofing	
_____	Footing forms, lumber for footings and top of walls	$_____
_____	Rebar (reinforcing steel bars)	_____
_____	Concrete for footings, telepost pads, posts	_____
_____	Concrete walls	_____
_____	Pump charge (or charge for placing concrete)	_____
_____	Concrete additives (calcium and heat if req'd)	_____

WORKSHEET # 20 - Continued

ITEM CHECK

_____	Dampproofing, foundation waterproofing	$_____________
_____	Weeping tile around foundation walls	_____________
_____	Labor charge for laying weeping tile	_____________
_____	Washed gravel for covering weeping tile	_____________
_____	Other	_____________
	SUBTOTAL	**$**___________

3. Alternative Preserved wood basement

_____	Engineered drawings or engineered stamped	_____________
_____	Footing forms	_____________
_____	Concrete and reinforcing (if designed with concrete)	_____________
_____	Washed gravel delivered and placed	_____________
_____	Compacting gravel	_____________
_____	Preserved wood basement materials	_____________
_____	Subfloor materials	_____________
_____	Weeping tile, extra gravel, sump box	_____________
_____	Framing labor to construct basement	_____________
_____	Other	_____________
	SUBTOTAL	**$**___________

4. Basement Labor

_____	Crib footings	$_____________
_____	Crib walls (installing necessary basement beams)	_____________
_____	Supervise the pouring of concrete	_____________
_____	Install subfloor (if contracted by cribber)	_____________
_____	Other	_____________
	SUBTOTAL	**$**___________

5. Basement floor

_____	Basement gravel (should be clean 3/4" rock)	$_____________
_____	Preparation, placing gravel, levelling, tamping, etc.	_____________
_____	Concrete for basement floor	_____________
_____	Finishing labor	_____________
_____	Pit run or gravel for garage	_____________
_____	Preparation for garage floor (levelling, rebar, etc.)	_____________
_____	Concrete for garage floor (includes finishing)	_____________
_____	Other	_____________
	SUBTOTAL	**$**___________

STRUCTURE

ITEM CHECK

6. Framing materials

(checklist of main sections from material quantity take-off)

Check	Item	Amount
_____	Main subfloor	$_____
_____	Glue, nails, insulation stops, other odds and ends	_____
_____	Walls	_____
_____	Second story subfloor	_____
_____	Walls	_____
_____	Roof trusses	_____
_____	Roof materials	_____
_____	Pre-manufactured stairs and other stair materials	_____
_____	Garage walls	_____
_____	Garage roof	_____
_____	Basement framing materials	_____
_____	Decks, railings, stairs, bolts, finishing nails, etc.	_____
_____	Other	_____
	SUBTOTAL	**$_____**

7. Framing Labor

Check	Item	Amount
_____	Contract or tender (check items listed below)	$_____
_____	Subfloor	_____
_____	Main Structure	_____
_____	Garage	_____
_____	Deck	_____
_____	Basement framing	_____
_____	Stairs (install factory manufactured stairs)	_____
_____	Extra items (see contract form for list of extras)	_____
_____	Other	_____
	SUBTOTAL	**$_____**

8. Doors and Windows

Check	Item	Amount
_____	Basement windows	$_____
_____	Doors and Windows	_____
_____	Window extras (grills, cladding, increased R value)	_____
_____	Garage door supplied and installed	_____
_____	Garage door opener supplied and installed	_____
_____	Skylights and stormdoors	_____
_____	Other	_____
_____	Other	_____
	SUBTOTAL	**$_____**

WORKSHEET # 20 - Continued

ITEM CHECK

9. Roofing

_____	Roofing Estimate	$__________
_____	Roofing materials (shakes, tiles, shingles etc.)	__________
_____	Roofing Labor	__________
_____	Accessories (flashing, nails, poly, glue, etc.)	__________
_____	Other	__________
	SUBTOTAL	**$**__________

10. Siding or Stucco and Brick Exteriors

_____	Siding Estimate	$__________
_____	Siding materials	__________
_____	Siding labor	__________
_____	Trim boards	__________
_____	Staining siding and / or trim	__________
_____	Stucco (paper, wire, scratch coat, finish coats)	__________
_____	Masonry - brick material and all supplies	__________
_____	Masonry - labor	__________
_____	Parging (finishing on exterior basement walls)	__________
_____	Other	__________
	SUBTOTAL	**$**__________

11. Fascia, Soffits, and Eavestroughing

_____	Soffit material	$__________
_____	Fascia material	__________
_____	Labor	__________
_____	Eavestroughing (material and installation)	__________
_____	Other	__________
	SUBTOTAL	**$**__________

MECHANICAL

12. Plumbing

_____	Plumbing contract including fixtures	$__________
_____	Fixtures supplied by builder (i.e. jacuzzi)	__________
_____	Gas line connection	__________
_____	Water and sewer connections	__________
_____	Other	__________
_____	Other	__________
	SUBTOTAL	**$**__________

WORKSHEET # 20 - Continued

ITEM CHECK

13. Heating

_____	Heating contract	$_________
_____	Heating extras (air cleaner, humidifier, etc.)	_________
_____	Other	_________
	SUBTOTAL	**$**_________

14. Electrical

_____	Electrical contract	$_________
_____	Electrical permits	_________
_____	Temporary power supply (saw service)	_________
_____	Service from main line to house	_________
_____	Fixtures (other than included in fixtures # 20)	_________
_____	Electrical trench - include winter allowance ?	_________
_____	Other	_________
	SUBTOTAL	**$**_________

FINISHING

15. Insulation and vapour barrier

_____	Wall insulation - main structure	$_________
_____	Basement wall insulation (inc. any rigid insulation)	_________
_____	Garage insulation	_________
_____	Roof insulation -batt	_________
_____	Roof insulation - Blown in	_________
_____	Caulking	_________
_____	Poly vapour barrier and installation	_________
_____	Labor for caulking and installing poly	_________
_____	Other	_________
	SUBTOTAL	**$**_________

16. Drywall

_____	Drywall contract	$_________
_____	Drywall material	_________
_____	Boarding labor	_________
_____	Taping and sanding	_________
_____	Texturing ceilings	_________
_____	Garage drywall	_________
_____	Other	_________
	SUBTOTAL	**$**_________

WORKSHEET # 20 - Continued

ITEM CHECK

17. Kitchen cabinets, vanities

_____	Materials - cabinets and vanities	$__________
_____	Laundry room cabinets	__________
_____	Countertops	__________
_____	Marble countertops	__________
_____	Installation	__________
_____	Medicine cabinets	__________
_____	Extras (pull out drawers, counter saver, clock etc.)	__________
_____	Other	__________
_____	Other	__________
	SUBTOTAL	**$**__________

18. Floor coverings

_____	Carpet and lino	$__________
_____	Underpad	__________
_____	Installation	__________
_____	Extras, metal strips, stairs, yardage allowance	__________
_____	Hardwood flooring - material and labor	__________
_____	Floor tiles, materials and labor	__________
_____	Other	__________
	SUBTOTAL	**$**__________

19. Ceramic tiles, bathroom accessories

_____	Tiles - materials and labor for bathrooms	$__________
_____	Tiles - kitchen backsplash	__________
_____	Mirrors	__________
_____	Bathroom accessories (towel bars, paper roller,etc.)	__________
_____	Marble trim around jacuzzi tub, shower door	__________
_____	Other	__________
	SUBTOTAL	**$**__________

20. Electrical light fixtures

_____	Light fixtures and bulbs (incl. fans, door chimes)	$__________
_____	Installation (included in electrical contract)	__________
_____	Extra indirect lighting (pot lights, flourescents,etc.)	__________
_____	Sunshine ceiling in kitchen (Fixtures and panels)	__________
_____	Extra wiring and plugs	__________
_____	Other	__________
	SUBTOTAL	**$**__________

WORKSHEET # 20 - Continued

ITEM CHECK		
	21. Painting	
_____	Painting contract, brush and roll primer and finish	$__________
_____	Primer and paint exterior (if not incl. in contract)	__________
_____	Painting exterior windows	__________
_____	Primer and paint interior (materials only)	__________
_____	Thinner, brushes, rollers, trays, etc.	__________
_____	Painting interior trim, spray staining - materials	__________
_____	Labor contract for staining and lacquering	__________
_____	Paint basement floor and steps	__________
_____	Paint garage door, front door	__________
_____	Other	__________
	SUBTOTAL	**$**__________
	22. Interior finishing	
_____	Finishing materials (baseboards, casings, mouldings)	$__________
_____	Railings - material and installation	__________
_____	Shelving materials, underlay	__________
_____	Extras (feature walls, ceiling trim, book cases)	__________
_____	Interior doors, bi-folds, bi-pass, pocket or french	__________
_____	Finishing around fireplace - materials	__________
_____	Hardware, door locks, closet rods, etc.	__________
_____	Mirror bi-pass doors	__________
_____	Finishing labor	__________
_____	other	__________
	SUBTOTAL	**$**__________
	23. Parging	
_____	Material and labor to finish exterior basement walls	$__________
	SUBTOTAL	**$**__________
	24. Steps, sidewalks, driveway	
_____	Steps - front and back	$__________
_____	Railings (if required)	__________
_____	Sidewalk blocks or poured, placed and finished	__________
_____	Concrete driveway (prep, placed and finished)	__________
_____	Gravel for driveway	__________
_____	Reinforcing wire, rebar, supports or grade beam	__________
_____	Retaining wall and other miscellaneous items	__________
	SUBTOTAL	**$**__________

WORKSHEET # 20 - Continued

ITEM CHECK

25. Extras

_____	Damage deposit to developer	$__________
_____	Financing land (accrued interest)	__________
_____	Land taxes	__________
_____	Mortgage application and appraisal	__________
_____	Construction insurance	__________
_____	Rentals (gravel vibrator, jack hammer, water pump, etc.)	__________
_____	Heating fuel, power	__________
_____	Appliances, fridge, stove, dishwasher, micro.	__________
_____	Extra appliances, washer, dryer, garberator, etc.	__________
_____	Fireplaces, mantel, brick, hearth, labor	__________
_____	Exterior basement entry, other construction extras	__________
_____	Built-in or rough-in vacuum system	__________
_____	Wallpaper and other decorating	__________
_____	Curtains, blinds and other window coverings	__________
_____	Worker's compensation (if trades not covered)	__________
_____	Delivery costs	__________
_____	Decks, patio, fence	__________
_____	Sod, trees, shrubs	__________
_____	Garbage removal and dumping charges	__________
_____	Tools, skill saw, hammer, wheelbarrel, etc.	__________
_____	Basement development	__________
_____	Other	__________
	SUBTOTAL	**$**__________

26. Services and permits

_____	Building permit (development permit)	$__________
_____	Water meter	__________
_____	Natural gas application	__________
_____	Nat. gas trenching and installation of service	__________
_____	Other trenching, plumbing, electrical	__________
_____	Electrical permit (see electrical contract)	__________
_____	Sandfill over services if backfill is frozen chunks	__________
_____	Winter allowance for trenching and excavation	__________
_____	Telephone installation fees	__________
_____	Cable television installation fees	__________
_____	Septic system, tile bed, water well (if required)	__________
_____	Other	__________
	SUBTOTAL	**$**__________

WORKSHEET # 20 - Continued

ITEM CHECK

27. Builder's costs (if required)

Check	Item	Amount
_____	Salesperson commissions, salaries	$_____________
_____	Displays and advertising	_____________
_____	Indirect building costs, depreciation, tools,	_____________
_____	Auto, truck, gas, maintenance, insurance	_____________
_____	Site supervision, wages, UIC, vacation pay, etc.	_____________
_____	Office and administrative, salaries, rent, utilities	_____________
_____	Office printing, stationary, legal, accounting	_____________
_____	Insurance, licences, dues, warranties	_____________
_____	Bank charges, interest, business taxes	_____________
_____	Profit	_____________
	SUBTOTAL	**$**___________

28. Legal and interest costs

Check	Item	Amount
_____	Legal fees	$_____________
_____	Land financing interest	_____________
_____	Mortgage draw interest	_____________
_____	Interim financing interest	_____________
_____	Service charges	_____________
_____	Credit financing interest (only if negotiated)	_____________
_____	Other	_____________
	SUBTOTAL	**$**___________

ESTIMATING SUMMARY

First floor area	__________	**Cost / sq. ft.**
Second floor area	__________	**Cost / sq. ft.**
Other	__________	**Cost / sq. ft.**
Total area developed	__________	**Cost / sq. ft.**

Estimated total construction costs **$** ___________

Add contingency factor, legal and interest **$** ___________

ESTIMATED TOTAL CONSTRUCTION COSTS **$** ___________

COST SUMMARY AND LOAN CALCULATION

Builders Name(s)

____

Preparation

- Blueprints, site plans, survey, etc.	(1) $	________

Foundation

- Excavating, backfill, trenching, grading, loaming	(2) $	________
- Basement materials (forms, concrete footings, concrete/preserved wood walls, weeping tile, dampproofing)	(3) $	________
- Basement labour, cribbing concrete or framing walls	(4) $	________
- Basement floor (gravel, concrete and finishing labour)	(5) $	________

Structure

- Framing materials (floor joists, sheathing, wall studs, roof framing, trusses, etc.)	(6) $	________
- Framing labour	(7) $	________
- Doors and windows	(8) $	________
- Roofing material and labour	(9) $	________
- Siding, brick and/or stucco, battons, trims, etc.	(10) $	________
- Fascia, soffits and eavestroughing	(11) $	________

Mechanical (all items supply and install)

- Plumbing	(12) $	________
- Heating	(13) $	________
- Electrical	(14) $	________

Finishing (all items summary of material and labour)

- Insulation, caulking and poly vapour barrier	(15) $	________
- Drywall (material and boarding, taping, sanding)	(16) $	________
- Kitchen cabinets, bath and laundry room vanities and counter tops	(17) $	________
- Floor coverings (underlay, carpet, linoleum, tile)	(18) $	________
- Ceramic tiles, Bathroom accessories	(19) $	________
- Electrical fixtures (all lights, bulbs, luminous panels)	(20) $	________
- Painting (primer, paint, stain, lacquer)	(21) $	________
- Interior finishing (railings, casings, baseboards, hardware)(labour)	(22) $	________
- Exterior parging	(23) $	________
- Steps, sidewalks, garage pad and driveway	(24) $	________
- Extras (i.e. appliances, fireplace, garage door, etc.)	(25) $	________

Services and permits	(26) $	________
Builders costs (general contractors or consultant fees)	(27) $	________

CONSTRUCTION COSTS ________

Contingency Factor (3-5%)	$	________
Land (incl. interest, taxes, security deposit)	$	________
TOTAL (finished home and land)	$	________
Legal and interest costs (2-4% above total)	(28) $	________
SUB TOTAL	$	________
Mortgage insurance fee, appraisal fee	$	________
Other	$	________
TOTAL CONSTRUCTION COST	**$**	________
LESS: Downpayment (deposits paid, items purchased, cash, other equity)	$	________
TOTAL LOAN / MORTGAGE REQUIRED	**$**	________

METRIC CONVERSIONS

LENGTH

1 mm (millimetre) = .03937 inch	1 inch = 25.4 mm (exactly)
1 m (metre) = 3.28084 feet	1 foot = 0.3048 m (exactly)
1 m (metre) = 1.09361 yards	1 yard = 0.9144 m (exactly)
1 km (kilometre) = 0.621371 mile	1 mile = 1.60934 km

AREA

1 cm^2 (square centimetre) = 0.1550 square inch	1 square inch = 6.4516 cm^2
1 m^2 (square metre) = 10.7639 square feet	1 square foot = 0.0929030 m^2
1 m^2 (square metre) = 1.19599 square yards	1 square yard = 0.836127 m^2
1 ha (hectare) = 2.47105 acres	1 acre = 0.404686 ha

VOLUME

1m^3 (cubic metre) = 35.3147 cubic feet	1 cubic foot = 0.0283168 m^3
1m^3 (cubic metre) = 1.30795 cubic yards	1 cubic foot = 0.764555 m^3
1 L (litre) = 0.219969 gallon	1 gallon = 4.54609 L (exactly)

MASS

1 kg (kilogram) = 2.20462 pounds	1 pound = 0.453592 kg

Concrete

psi	MPa
2175	15
2500	17.5
2900	20.0
3625	25.0
3987	27.5

Rebar

#2 - 6 mm
#4 - 10 mm
3" telepost = 75mm

2" - 38 mm	5" - 114 mm	11" - 260 mm
2 1/2 " - 51 mm	6" - 140 mm	12" - 286 mm
3" - 64 mm	7" - 165 mm	14" - 337 mm
3 1/2" - 76 mm	8" - 184 mm	16" - 387 mm
4" - 89 mm	9" - 210 mm	
4 1/2" - 102 mm	10" - 235 mm	

Note: Standard metric stud length is 2310 mm, with double top plate and single bottom plate for 2400 mm ceiling height.

Insulation	Polythene Vapour Barrier Thickness	Gypsum Board
R-8 - RSI 1.4	2 mil	1/2" - 12.7 mm
R-12 - RSI 2.1	4 mil	
R-14 - RSI 2.5		5/8" - 15.9 mm
R-20 - RSI 3.2	6mil	
R-28 - RSI 4.8		
R-32 - RSI 5.3		

Doors:

3'0" = 914 mm
2'8" = 812 mm
2'6" = 762 mm
2'4" = 711 mm
2'0" = 609 mm
6'8" = 2032 mm

210 lb. shingles are referred to as #10 standard
231 lb. shingles are referred to as "low slope"

MATERIALS LIST FOR SUPPLIERS

		FOOTINGS			
Footing Forms		Lineal Feet - 2 x 6 Random Length Spruce			
Footing Forms		LFT - R/L Spruce			
Stake Material		Pieces - Common Spruce			
Reinforcing		LFT - 20' Rebar			
Drainage		LFT -4" Weeping Tile			
Drainage		Pieces of 4" Joiners			
Nails		LBS - Coated Nails			
Bolts		Pieces - Anchor Bolts			
Piers		Sono Tube			
Misc.					
		FOOTINGS TOTAL			
		CONCRETE WALLS / PONYWALLS			
Ladder, Top of Concrete Wall		LFT - 2 x 4 - R/L Construction Spruce			
Window Bucks (Frame)		LFT - 2 x 10 - R/L Fir			
Bracing		LFT - 2 x 4 - R/L Construction Spruce			
Sill Plates		LFT - 2 x 8 - R/L Spruce or Fir			
Studs		Pieces 2 x 6 Construction Spruce			
Plates		LFT 2 x 6 - R/L Construction Spruce			
Headers		LFT 2 x 10 - R/L Fir			
Exterior Sheathing		Pieces 4 x 8 OSB Board or Plywood			
Nails		Coated			
		CONCRETE WALLS AND PONY WALLS TOTAL			$

TOTALS THIS PAGE $__________

MATERIALS LIST FOR SUPPLIERS

		ALTERNATE - PRESERVED WOOD FOUNDATION			
Footing Plate		LFT 2 x 8 Random Length P.W.F.			
Footing Plate		LFT 2 x 10 R/L P.W.F.			
Studs		Pieces 2 x 4 - 8 Ft. P.W.F.			
Studs		Pieces 2 x 6 - 8 Ft. P.W.F.			
Studs		Pieces 2 x 8 - 8 Ft. P.W.F.			
Bottom Plates		LFT 2 x 6 - R/L P.W.F.			
Top Plates		LFT 2 x 6 - R/L P.W.F.			
Bottom Plates		LFT 2 x 4 - R/L P.W.F.			
Top Plates		LFT 2 x 4 - R/L P.W.F.			
Headers		LFT 2 x 10 - R/L Fir			
Screed Board		LFT 1 x 3 - R/L P.W.F.			
Exterior Sheating		Pieces 4 x 8 - 1/2" P.W.F. Plywood			
Exterior Sheating		Pieces 4 x 8 - 5/8" P.W.F. Plywood			
Blocking		LFT 2 x 4 - R/L P.W.F.			
Grade Board		Pieces 4 x 8 - 1/2" P.W.F. Plywood Cut 12" Wide			
Corner Board		Pieces 4 x 8 - 1/2" P.W.F. Plywood Cut 8" Wide			
Vapour Barrier		Rolls 10" - 6 Mil Poly			
Tar Coating		Gallons P.W.F. Tank Solution			
Caulking		Tubes Butyl Caulking			
Nails		Galvanized 3-1/2"			
Nails		Galvanized 2"			
P.W.F. for on site cuts		1 Gallon Penta Green			
Misc.					
		P.W.F. Basement Total			

TOTALS THIS PAGE $______

MATERIALS LIST FOR SUPPLIERS

		SUBFLOOR			
Posts		Adjustable Teleposts			
Posts		Pieces Fir Posts			
Beam(s) & Floor Joists		8 Ft.___ 10 Ft.___ 12 Ft.___ 14 Ft.___ 16 Ft.___			
		18 Ft.___ 20 Ft.___ 22 Ft.___ 24 Ft.___ Fir 2x10			
Beam(s) & Floor Joists		8 Ft.___ 10 Ft.___ 12 Ft.___ 14 Ft.___ 16 Ft.___			
		18 Ft.___ 20 Ft.___ 22 Ft.___ 24 Ft.___ Fir 2x12			
Optional - Floor Trusses		TJI 9-3/8 or 11-7/8 or other Manuf. Specs			
Blocking & Trimmers		LFT 2 x ______ - R/L			
Headers		LFT 2 x 10 - R/L Fir			
Bridging		Pieces 2 x 2 for 16" on center			
Bridging		Pieces 2 x 2 for 12" on center			
Joist Hangers		Pieces - Single Joist Hangers			
Joist Hangers		Pieces - Double Joist Hangers			
Floor Sheathing		Pieces - 4 x 8 OSB Board or Plywood			
Floor Adhesive (Glue)		Tubes PL 400 Floor Blue			
Insulation		One Bag R - 20 Fiberglass Batt			
Other		TOTAL			
		MAIN FLOOR FRAMING			
Studs Interior		Pieces - 2 x 4 Construction Spruce			
Studs Exterior Walls		Pieces - 2 x 6 Construction Spruce			
Studs For High Ceilings		Pieces - 2 x - 10 Ft.___ 12 Ft.___ 14 Ft.___ 16 Ft.___			
Plates Interior		LFT 2 x 4 - R/L Construction Spruce			
Plates Exterior		LFT 2 x 6 - R/L Construction Spruce			

TOTALS THIS PAGE $__________

MATERIALS LIST FOR SUPPLIERS

Headers		LFT 2 x 10 Fir			
Exterior Sheathing		Pieces 4 x 8 OSB Board or Plywood			
Misc. Framing, Bracing		LFT 2 x 4 - R/L Construction Spruce			
Beam(s)		Microlam Laminated Plywood			
Beam(s)		Microlam Laminated Plywood			
Vapour Barrier		Rolls ____ Mil Poly for Framing			
Vapour Barrier		Rolls - Black Building Paper			
Insulation		Bundles 15" Fiberglass (For Framing Only)			
Pocket Door		Pocket Door Framing & Hardware			
Misc.					
Misc.					
		MAIN FLOOR FRAMING TOTAL			
		2ND FLOOR SUBFLOOR & FRAMING			
Beams - Fir		LFT 2 x ___; 8 Ft.___ 10 Ft.___ 12 Ft.___			
		14 Ft.___ 16 Ft.___ 18 Ft.___			
Josits - Fir		LFT 2 x ___; 8 Ft.___ 10 Ft.___ 12 Ft.___			
		14 Ft.___ 16 Ft.___ 18 Ft.___			
		20 Ft.___ 22 Ft.___ 24 Ft.___			
Joists (Floor Trusses)		Alternative Floor System Specified By Supplier			
Trimmers		LFT 2 x ___ R/L			
Bridging		Pieces 2 x 2 Bridging (16" on Center)			
Bridging		Pieces 2 x 2 Bridging (12" on Center)			
Joist Hangers		Pieces Single Joist Hangers			
Joist Hangers		Pieces Double Joist Hangers			
Floor Sheathing		4 x 8 OSB Board or Plywood Sheathing			
Glue		Tubes PL 400 Floor Glue			
Studs Interior		Pieces 2 x 4 Construction Spruce			

TOTALS THIS PAGE $___________

MATERIALS LIST FOR SUPPLIERS

Studs Exterior		Pieces 2 x 6 Construction Spruce			
Plates Interior		LFT 2 x 4 R/L Construction Spruce			
Plates Exterior		LFT 2 x 6 R/L Construction Spruce			
Headers		LFT 2 x 10; 8 Ft.___ 10 Ft.___ 12 Ft.___			
Exterior Sheathing		Pieces 4 x 8 3/8" OSB Board or Spruce Ply			
Blocking		LFT 2 x 4 R/L Construction Spruce			
Misc.					
Misc.					
		2ND FLOOR SUBFLOOR & FRAMING TOTAL			
		GARAGE AND ROOF FRAMING			
Studs		Pieces 2 x ___ Construction Spruce			
Plates		LFT 2 x ___ R/L Construction Spruce			
Header		LFT 2 x 10; 8 Ft.___ 10 Ft.___ 12 Ft.___			
Garage Door Header		LFT 2 x ___ 18' or Mfg. Truss Design			
Exterior Sheathing		4 x 8 3/8" OSB Board or 3/8" Spruce Plywood			
Roof Trusses		As On Plans Specified By Truss Mfg.			
Gable Sheathing		4 x 8 3/8" OSB Board or 3/8" Spruce Plywood			
Peak and Ladders		LFT 2 x 4 R/L Construction Spruce			
Fascia		LFT 2 x ___ R/L Spruce			
Soffit Nailers		LFT 2 x 2 R/L Spruce			
Soffit					
Soffit					
Vents		Soffit Vents			
Garage Door					
Casing		Brick Moulding LFT			
Strapping		LFT 1 x 4 Spruce			

TOTALS THIS PAGE $__________

MATERIALS LIST FOR SUPPLIERS

Shingles		Bundles Asphalt Shingle Type ______________			
Starter Poly		Rolls 44 Inch 6 Mil Poly			
Roof Sheathing		Pieces 4 x 8 OSB Board or Spruce Plywood			
Clips		Roof Sheathing Clips			
Nails					
Stops		Pieces Wax Insulation Stops			
Misc.					
Misc.					
		GARAGE AND ROOF TOTAL			
		ROOF ELEVATION			
Trusses		As Required On Plans			
Peak and Ladders		Lineal Ft.			
Gable Sheathing		Pieces 4 x 8 OSB Board or 3/8" Spruce Ply			
Fascia Nailers		LFT 2 x ___ R/L Construction Spruce			
Soffit Nailers		LFT 2 x ___ R/L Construction Spruce			
Soffit					
Soffit					
Vents		Pieces Soffit Vents			
Misc. Framing		LFT 2 x 4 R/L Construction Spruce			
Strapping		LFT 1 x 4 R/L Spruce			
Roof Sheating		Pieces 4 x 8 OSB Board or Spruce Plywood ___			
Clips		Pieces Plywood Clips			
Shingles (or other roofing)		Bundles Asphalt Shingles			
Starter Poly		Rolls 44 Inch 6 Mil Poly			
Nails					

TOTALS THIS PAGE $__________

MATERIALS LIST FOR SUPPLIERS

Nails					
Insulation Stops		Pieces Waxed Insulation Stops			
Fascia		Lineal FT. R/L _____ x _____			
Misc.					
Misc.					
		ROOF ELEVATION TOTAL			
		LANDING, STAIR & BASEMENT FRAMING			
Studs		Pieces			
Studs		Pieces			
Plates		Lineal FT			
Plates		Lineal FT			
Landing Joists		Lineal FT 2 x ___; 8 FT___ 10 FT___ 12 FT___			
Floor		Pieces 4 x 8 T&G OSB Board or Plywood			
Glue		Tubes PL 400 Floor Glue			
Stairs Pre-Manufactured		As Supplied By Manufacturer			
Stair Stringer		Lineal FT _____ No.1 Fir			
Stair Tread		Lineal FT 2 x _____ No.1 Fir			
Stair Tread		Pieces 4 x 8 - 1 Inch OSB Board or Fir Plywood			
Stair Risers		Pieces 4 x 8 - 1/2 Inch OSB Board or Spruce Ply			
Misc.					
Misc.					
		TOTAL LANDING AND STAIRS			

TOTALS THIS PAGE $__________

MATERIALS LIST FOR SUPPLIERS

		DRYWALL AND INSULATION DRYWALL SIZES			
Master Bedroom & Bath		(8) ____ (10) ____ (12) ____ (14) ____			
Bedroom #2		(8) ____ (10) ____ (12) ____ (14) ____			
Bedroom #3		(8) ____ (10) ____ (12) ____ (14) ____			
Bedroom #4		(8) ____ (10) ____ (12) ____ (14) ____			
Hallway		(8) ____ (10) ____ (12) ____ (14) ____			
Living Room		(8) ____ (10) ____ (12) ____ (14) ____			
Family Room		(8) ____ (10) ____ (12) ____ (14) ____			
Dining Room		(8) ____ (10) ____ (12) ____ (14) ____			
Kitchen		(8) ____ (10) ____ (12) ____ (14) ____			
Hallway		(8) ____ (10) ____ (12) ____ (14) ____			
Foyer		(8) ____ (10) ____ (12) ____ (14) ____			
Den		(8) ____ (10) ____ (12) ____ (14) ____			
Other		(8) ____ (10) ____ (12) ____ (14) ____			
Other		(8) ____ (10) ____ (12) ____ (14) ____			
TOTAL NO. OF SHEETS		(8) ____ (10) ____ (12) ____ (14) ____			
		DRYWALL TOTAL			
		INSULATION - DESCRIPTION			
Wall Insulation		(R-20)			
Basement Wall Insulation		(R-12) (R-20 for PWF Bsmt.)			
Ceiling Insulation		(R-40)			
Ceiling Insulation		(R-35) or other			

TOTALS THIS PAGE $________

MATERIALS LIST FOR SUPPLIERS

Vapour Barrier		No. of Rolls of 6 Mil Poly			
Caulking		No. of Tubes of Acoustical Sealant			
Staples		Boxes of Staples			
Misc.					
Misc.					
Misc.					
		INSULATION TOTAL			
		SIDING AND TRIMS			
Type					
Type					
Building Paper		Rolls Black Building Paper			
Trim		Lineal Feet			
Trim		Lineal Feet			
Stucco Wire					
Wire Nails (Install Only)					
Power Staples					
Misc.					
Misc.					
Misc.					
Misc.					
		SIDING AND TRIMS TOTAL			

TOTALS THIS PAGE $__________

MATERIALS LIST FOR SUPPLIERS

		DECKS			
Beam		Lineal FT			
Joists		Lineal FT 2 x ___ Sizes _____			
Joists		Lineal FT 2 x ___ Sizes _____			
Trimmer Joists		Lineal FT 2 x ___ R/L			
Floor					
Posts		Pieces			
Posts		Pieces			
Top Rail		Lineal Feet			
Bottom Rail		Lineal Feet			
Pickets		Pieces			
Nails for Joists					
Nails (screws) - Top of Deck					
Lag Bolts					
Misc.					
		DECKS TOTAL			
		FINISHING MATERIALS *DOOR SIZES AND DESCRIPTION*			
Door					
Door					
Door					
Door					

TOTALS THIS PAGE $______

MATERIALS LIST FOR SUPPLIERS

Item		Unit			
Casing Windows		Lineal Feet			
Casing Doors		Lineal Feet			
Pocket Door					
Closet Door					
Closet Door					
Closet Door					
Closet Door					
Closet Door					
Jambs					
Closet Door Casings		Lineal Feet			
Baseboards		Lineal Feet			
Attic Trim		Lineal Feet			
Stair Trim		Lineal Feet			
Shelving					
Shelf Support		Pieces or Lineal Feet			
Misc.					
Handrail (top Rail)		Lineal Feet			
Posts		Pieces			
Spindles		Pieces			
Railing Base		Lineal Feet			
Misc. Railing Parts					
Other Railing - Handrail		Lineal Feet			
Railing Hardware					
Underlay		Pieces 4 x 8 3/8 inch K-3 Board			
Misc.					
		FINISHING MATERIAL TOTAL			

TOTALS THIS PAGE $________

MATERIALS LIST FOR SUPPLIERS

		HARDWARE			
Front Door Set		Decorative Door Handle			
Door Knobs		Passage			
Door Knobs		Privacy			
Door Knobs		Lock Sets			
Door Knobs		Dead Bolts			
Door Knobs		Pocket Door Sets			
Misc.					
Rods		Closet Rods - Length _____			
Rods		Closet Rods - Length _____			
Rods		Closet Rods - Length _____			
Rods		Closet Rods - Length _____			
Shelf Supports					
Door Stops		Spring Type			
Door Stops		Floor Type			
Door Stops & Hardware		French Doors			
Handrail		Handrail Brackets, Screws			
Misc.					

TOTALS THIS PAGE $_______________

MATERIALS LIST FOR SUPPLIERS

		HARDWARE TOTAL			
		NAILS			
Basement					
Subfloor Joists & Sheathing					
Framing and Wall Sheathing					
Roof Sheathing					
Drywall Screws					
Siding					
Deck Joists & Deck Top					
Finishing Nails					
Finishing Nails					
			NAILS	TOTAL	$

TOTAL PAGE ONE		
PAGE TWO		
PAGE THREE		
PAGE FOUR		
PAGE FIVE		
PAGE SIX		
PAGE SEVEN		
PAGE EIGHT		
PAGE NINE		
PAGE TEN		
PAGE ELEVEN		
PAGE TWELVE		
PAGE THIRTEEN		
GRAND TOTAL PAGES 1 TO 13		$

TOTALS THIS PAGE $________

12

YOUR BUYING CONTRACTING GUIDE

Researching the Market

It pays to compare several estimates! Most suppliers have different overhead costs, different buying power, and different qualifications, which in turn affects their cost and the prices they charge you. Comparing **only** price can be a costly mistake in the long run. There are many other factors to consider, such as quality, payment terms, time required to complete the job, guarantees, service, etc. This information can be obtained only by thoroughly researching the market.

How do you save time? How can you be efficient and effective in conducting your research? You should get at least 3 estimates for each of the 18 major jobs or approximately 54 estimates. In many towns you will not have this many suppliers and trades available; however, you should research as many as possible. If you have 5 sets of plans to send out and it takes about a week to receive each estimate, it will take 11 weeks just to price out the home, and you will be running around constantly.

If you have 20 sets of plans, you can completely cost out your home in less than 3 weeks. The cost for the extra 15 sets of plans (15 x $5 = $75) is insignificant in comparison to the potential savings in time and money.

Know what you want and need before looking. Be as clear as possible. Most complaints during construction stem from a lack of communication. The job wasn't explained clearly enough. You invite a painter to paint your room green, but you don't specify what shade. Later you have a battle because the job isn't done as anticipated. You can only blame yourself for not giving specific directions in the beginning.

When you ask for estimates, it is essential for comparison purposes that you give all trades and suppliers the same information. For example, you request estimates for your lumber from three separate suppliers. If you are not specific about what you want, you will receive three completely different amounts of materials, grades of materials, payment terms, etc. This information will be very difficult to compare.

On the other hand, if you do your own material quantity take-off and present your supplier with a list of materials and grades (e.g. No. 1 fir, construction grade spruce), you will end up with prices you can compare. If you are reluctant to do a material take-off, you can still do a dollar for dollar comparison by making up a list of products and taking it to various suppliers. Present them with your plans and ask their estimator to provide you with detailed prices. The one with the best price, service, and payment terms will receive your business.

Checking Qualifications and Reputation

Check each trade's qualifications and reputation. Find out the answers to these questions:

- How long have they been in business?
- What is the background of the company?
- What is the company's financial status?
- How much time do they spend on the job?
- What is their reputation from referrals?
- Will they guarantee their work and honour service?
- Is it their policy to give refunds or exchanges or to cancel a contract?
- Where was the last job they did? How long ago?
- Are they doing a job now? Where is it? (Go to the jobsite and talk to the owner or workers.)

You should also talk to previous customers. Ask them:

- Does the contractor stay with the job day after day until completed?
- Do you have to keep phoning to get the job finished?
- How is the workmanship or craftsmanship?

Other places to check are the Better Business Bureau or contractors' and trade associations (e.g. Association of Insulation Installers).

In most provinces and states; electricians, gas fitters and plumbers **must be certified.** They will carry a journeyman pocket identification showing their certification. The pocket card may also show completion of an apprenticeship program. The journeyman certificate proves that they have a certain minimum number of years of training in the field. The apprenticeship certificate is a step above and shows actual written qualifications as well as experience in the field. You can ask to see these qualifications.

All trades should have a provincial business licence from their consumer and corporate affairs department to deal directly with a consumer. If they are licensed, they are eligible for a bond so that if they start work on your place and don't complete it you can claim against their bond.

When they are operating without a licence, they can't be bonded and you have no recourse as a consumer. There are a number of contractors who don't have licences and who don't want licences. Many who don't have licences do excellent work, but you take a risk. The main thing to do is check with previous customers and make sure that those who are required to have certification have it.

Obtaining an Estimate and a Contract

The estimate you receive will be the basis for drawing up a contract. You will find varying opinions regarding what information an estimate should include and what a contract should include.

In most cases, when an owner-builder receives an estimate, the work is done based on that estimate and no other contract is formed. Too often , not enough information is specified in writing, which leads to problems down the road. The best method is to form a separate contract, but this results in a great deal of work when you are contracting up to 20 suppliers and trades. One alternative is to ensure that all the required details are specified in the estimate. When the estimate is accepted and signed by both parties, it becomes the contract. Another alternative is to attach an additional sheet of conditions and terms and have both copies signed by both parties at the time the estimate is accepted. See the examples of contracts included in this chapter for your use.
Since the estimate will either be the basis for or eventually become the contract, it should specify the following:

- The deadline date: The date on which the estimate must be submitted to be of any consideration to the owner. You may not be in a rush, but if you don't specify a date, you may never get your plans back and it will take months to price out your home.
- The current date!
- The name and address of the individual or company
- The name and address of the owner and address of the job site
- The lead time required: How much notice is required before they start the job?
- A full description of the work to be done: Do not leave anything to be misunderstood. Put it in writing and you will have no problems later on.
- A full description of the materials to be used and their quality: The description should mention that the products or equipment are new. Include name and model number (if necessary).
- A clear explanation of any guaranty or warranties that are offered: This must be specific, e.g. lifetime warranty - who's life? The Sale of Goods Act implies that the goods must be serviceable for what you bought them for. It implies there is a warranty. Some warranties are worded to take away rights that you normally would have.
- The total price including material and labour and how long it is good for. The final price must be close to the contracted price.
- The method and terms of payment: When possible, try to set up credit financing where you will have up to 30 days to pay your account without paying interest.
- The amount of down payment: You should try not to place any money down until you have received something for it; however, sometimes you will have no choice. Don't give deposits a long time in advance of requiring the service. Regardless of a good sale price, you run the risk of losing your down payment or committing yourself too early and passing up other good opportunities.
- The schedule of payments: A schedule of payments should be detailed on the contract to avoid problems. A plumber may request 50% after rough-in and balance on completion. This is fine, but the plumbing should pass the required inspections first before he receives payment.

- A statement that the contractor or worker has the necessary insurance to protect the homeowner from any liability that may result from the work: This statement will protect the owner-builder in case someone is injured as a result of a trade's negligence, for example, if a framer sheets your roof and the next day someone falls through a rotten sheet of plywood.
- A statement that worker's compensation will be the responsibility of the contractor: As a safety precaution, you can obtain the worker's compensation number from any trade before he enters the job site. Next, phone the compensation board to see if they are in good standing.
- A statement that all necessary permits will be obtained and that all the work will be done in strict accordance with local building regulations and the specifications in the plans.
- A statement that the contractor will take all necessary precautions to protect the job site from damage during work and will repair or replace any property damaged through his or her fault.
- A statement that the contractor will be responsible for cleaning up and removing all debris from the site after the work has been completed.
- Signatures of both the homeowner and contractor including the date of signing.
- Other information: Leave room on the contract for other information. Don't forget anything. Anything you discuss should be put into the contract in writing.

When you make payments, cover yourself by keeping records. Don't pay by cash unless you receive a signed written receipt. Another alternative is to buy a bank draft. The draft is as good as cash and you still have a cheque for your records as well as a bank draft receipt. For additional protection, mark on the invoice "PAID IN FULL by CASH or CHEQUE" and have it signed by the contractor. Always get as much protection as you can at the time. You may want to have a witness if one is available. Unless you have proof of payment, your supplier could go bankrupt and seek payment from you a second time because the auditors cannot find a record of payment.

Builder's or Mechanic's's Lien

When a person performs work or supplies materials in the construction of a building, it is physically impossible to retain possession. Under the law of real property, when goods are affixed permanently to land they become fixtures. The supplier is not permitted to take them from the property. The Mechanic's (or Builder's) Lien Act of each province provides protection by allowing creditors an interest in the land as security for payment. The act also provides protection for the builder against liens which could cause them to pay for goods or services twice. These provisions can be best understood by example.

Joe's Drywall was hired to drywall your home. The contract states he is to receive payment within 30 days of completion. The 30 days are up and you still haven't paid Joe. Joe has a certain amount of time after substantial completion of the contract to register a lien against your property. The act protects Joe.

Now, reverse the situation. How does the act protect you? Suppose you hire Joe to do the drywall and Joe hires a boarder, taper, and sander (A, B and C). You have no contract with these third parties and likewise no knowledge of how or when they are to be paid. They have a right to full payment for services to your property and have a certain amount of time after the work is complete to place a lien against your title.

You need to holdback 15% of the **total** contract price for a specific period of time. Unless you withheld 15% and searched the title for liens at the end of the time period, you can be held liable to pay the amount owed to A, B and C. Under the act, by holding back 15% you are only liable for payments to A, B and C for the amount of the 15% holdback.

(NOTE: As the percentage of contract price and amount of time varies among provinces, check your local regulations.)

The following questions and answers may give a better understanding of liens.

Q. What is a lien?
A. A lien is a charge registering an interest in the land as security for payment.
Q. How can I search my title?
A. A title search can be performed easily by identifying yourself and paying a small fee at the land titles office. You can easily hire your lawyer to do the checking for you.
Q. What do I do if I have a problem?
A. The first thing is to meet with your tradesman or supplier and talk out your differences. In many cases, the problem may only be a minor misunderstanding. Don't let the problem sit or assume it will work out in the long run. Approach it as soon as possible with an open mind. Avoid taking your case to court. Legal fees and court handling costs are not recoverable.
There may be some legislation that backs your claim. Go to the people who administer the legislation and research your position.
See your lawyer for advice.
You can settle your dispute in small claims court if the amount is under the limit for your province. The court clerk will give you the forms and help you fill them out.
Take precautions to prevent a lien. Suppose you were worried that a trade was in financial trouble and could not pay his supplier. You can provide some insurance that the supplier will receive payment by making the cheque payable jointly to both the tradesman and his supplier.
Q. What should I do if a lien is placed against my property?
A. Contact your lawyer. The amount of the lien or holdback will be paid into court. The judge will decide whether the lien claim is valid.
Q. Can a lien halt construction?
A. If you prefer to fight the lien rather than pay the money into court, the charge will remain against your title which will halt mortgage draws. When you run out of funds, your construction will stop until the lien is settled.
Q. Must I inform a trade in advance that I will be holding back a portion of the contract?

A. No. It is your legal right and it does not have to be in the contract. Every contractor should be aware of this. It is advisable, however, that you inform them of your policy regarding payments. Otherwise, when paying, send in your cheque stating that the holdback will be paid after a title search has been conducted on a certain date. The contractor will have received most of the payment. He will not file a lien against your property. The only person who would is a third party who has not been paid by your contractor.

Q. Is it necessary to hold back in every contract?

A. No. You are safe to pay the whole amount when there are no third parties involved and you are certain the trade is in good standing and will pay suppliers.

Q. When the payments are scheduled, do I hold back a portion of each payment?

A. The holdback must be calculated on the entire contract price. It can be taken off the last payment or if the last payment is not large enough, the percentage can be taken from each scheduled payment.

Q. When is a contract complete? What if there is one little thing left to do?

A. In the Builder's Lien Act, "completion of the contract" means substantial performance, not necessarily total performance of the contract. If the work is near completion and you encounter difficulty getting the contractor to finish the job, the contract may be deemed to be substantially performed.

Q. What policy should I follow regarding my changes or additional costs after a contract is signed?

A. The trade or supplier should receive your approval first. Both parties should initial the changes on the contract. If required, attach an additional sheet to both copies of the contract and note, continued on next page.

Q. What other protection do I have as a consumer?

A. The Unfair Trade Practices Act in most provinces aim to prevent unfair business acts or practices and aid consumers in recovering losses. This act protects consumers primarily against transactions where products are being misrepresented.

CONTRACT
EXCAVATION-BACKFILL-GRADING

DEADLINE DATE - TENDER MUST BE RETURNED BY________________

Plan no.__________

House Size__________

Builder

Name___________________________

Address_________________________

Phone__________________________

Contractor

Name__________________________

Address________________________

Phone_________________________

Job location____

Lead time required____

Estimated time required to complete the job_____________________________

Description of the job: (initial items included)

_____ **Excavate** basement to required elevation providing access ramps for materials and concrete trucks. The excavation will allow sufficient space around foundation for cribbers, weeping tile and dampproofing contractors.

_____ **Backfill** around basement to required elevation.

_____ **Rough Grade** the lot including hauling away fill or hauling in extra fill. The lot will be prepared ready for loam and rough graded for concrete drive.

_____ **Trenching** for water and sewer services and electrical underground.

_____ **Backfill** trenches as soon as possible following passed inspections of plumbing and electrical services.

*** Total price for excavation, trenching, backfill and grading $________

Rental rate for equipment:

$______________ per hour for approximately ______________ hour(s)

$______________ per hour for approximately ______________ hour(s)

$______________ per hour for approximately ______________ hour(s)

An estimated contract price will be reasonably close to the final price.

Additional charges:

Moving equipment: Excavation ______________

Backfill ______________

Grading ______________

Sand: $______________ per cu. yd. delivered

$______________ per cu. yd. delivered and placed into basement

3/4" Washed Rock: $__________ per cu. yd. delivered

$__________ per cu. yd. delivered and placed

Other ____

____

Terms of payment ____

____

*** This offer will remain in effect until ______________________________

***Conditions*:**

The contractor will not pile fill on adjacent lots.

The contractor will possess the necessary insurance to protect the builder from any liability which may result from his work.

The contractor will take all necessary precautions to protect the job site from damage and will repair or replace any property damaged through his fault including sidewalks and services.

All work will be done in strict accordance with local building regulations and the specifications in the plans.

Other: ____

CONTRACTOR ____

DATE OF OFFER ____

SIGNATURE ____

BUILDER ____

DATE OF ACCEPTANCE ____

SIGNATURE ____

The builder reserves the right to hold ______% of the total contract price for ______ days after completion.

CONTRACT
SUPPLY, DELIVER AND PLACE CONCRETE!

DEADLINE DATE - TENDER MUST BE RETURNED BY_______________

Plan no.__________
House Size__________

Builder
Name______________________________
Address______________________________
Phone______________________________

Contractor
Name______________________________
Address______________________________
Phone______________________________

Job location____
Lead time required____
Estimated time required to complete the job________________________________

Description: Supply, deliver and place concrete!

	TYPE (10)(50)	STRENGTH (MPA)	AMOUNT REQUIRED M^3	PRICE M^3	TOTAL
Footings					
Walls					
Basement Floor					
Driveway					
Other					

Additional charges:

Total Price to Supply Concrete $________________

Terms of payment ____
____
____

*** This offer will remain in effect until ____________________________

***Conditions*:**

The supplier will possess the necessary insurance to protect the builder from any liability which may result from the delivery or work.

The supplier will take all necessary precautions to protect the job site from damage and will be responsible to repair or replace any property damaged through his fault.

All materials supplied will be in accordance with local building regulations and specifications in the plans.

Other: ____

____

CONTRACTOR ____

DATE OF OFFER ____

SIGNATURE ____

BUILDER ____

DATE OF ACCEPTANCE ____

SIGNATURE ____

The builder reserves the right to hold ______% of the total contract price for ______ days after completion.

CONTRACT CRIBBING

DEADLINE DATE - TENDER MUST BE RETURNED BY________________

Plan no.__________

House Size__________

Builder
Name______________________________
Address____________________________
Phone______________________________

Contractor
Name____________________________
Address__________________________
Phone___________________________

Job location____
Lead time required____
Estimated time required to complete the job______________________________

Description of the job: (initial items included)

_____ Form concrete footings and pads
_____ Form walls
_____ Install basement window bucks
_____ Placement of reinforcing bars
_____ Install cast-in-place subfloor system (if specified)
_____ Supervise pouring of concrete footings
_____ Supervise pouring of concrete walls
_____ Remove debris and garbage on completion
_____ Other

**** TOTAL PRICE FOR CRIBBING FOOTINGS AND WALLS $____________**

Unit price for 8' foundation wall $_____________
(per lineal foot - not including floor joist system)

Unit price for 4' foundation wall $_____________
(per lineal foot)

Additional charges:

Charge for corners	$ _____________
Charge for basement entry	$ _____________
Place weeping tile and gravel	$ _____________
Seal snap tie holes	$ _____________
Install subfloor	$ _____________
Crib retaining walls	$ _____________
Other	$ _____________
TOTAL	**$**___________

Terms of payment ____
____

*** This offer will remain in effect until ____________________

***Conditi*ons:**

The contractor will possess the necessary insurance to protect the builder from any liability which may result from his work.

Worker's compensation will be the responsibility of the contractor.

The contractor will take all necessary precautions to protect the job site from damage and will repair or replace any property damaged through his fault.

All work will be done in strict accordance with local building regulations and the specifications in the plans.

Other: ____________________

CONTRACTOR ____________________
DATE OF OFFER ____________________

SIGNATURE ____________________

BUILDER ____________________
DATE OF ACCEPTANCE ____________________

SIGNATURE ____________________

The builder reserves the right to hold ______% of the total contract price for ______ days after completion.

CONTRACT
FRAMING WOOD BASEMENT AND SUBFLOOR

DEADLINE DATE - TENDER MUST BE RETURNED BY________________

Plan no.__________

House Size__________

Builder	***Contractor***
Name___________________________	Name___________________________
Address_________________________	Address_________________________
Phone__________________________	Phone__________________________

Job location____

Lead time required____

Estimated time required to complete the job___________________________

Foundation to be built on concrete footings ________ or washed rock _________

Description of the job: (initial items included)

_____ Check and ensure a level distribution of washed rock over building site for minimum depth of six inches and extented two feet outside foundation walls
_____ Install footing plates and level with transit
_____ Coat all wood cuts with (p.w.f.) preservative
_____ Use only hot dipped galvanized fasteners
_____ Install insulation where required during framing
_____ All below grade construction to be of preserved wood
_____ All joints in sheathing are to be sealed with approved caulking
_____ Apply exterior coating specially approved for preserved wood foundations
_____ Build sump box and cover
_____ Build telepost pads (if required)
_____ Construct and place wood beams, bearing walls and teleposts
_____ Install floor joists, crowns up with double joists under all parallel partitions (or install floor truss system as per truss design layout)
_____ Install trimmers, crossbridging, joist hangers, etc.
_____ Glue and securely nail (or screw if required) subfloor sheathing

Additional charges:

Level crushed rock	$ ____________
Crib concrete footings	$ ____________
Install subfloor	$ ____________
Supply nails, other materials	$ ____________
Supply all basement materials	$ ____________
Other	$ ____________
****TOTAL PRICE FOR FRAMING WOOD BASEMENT**	**$**____________

Terms of payment ______________________________________

____

*** This offer will remain in effect until ______________________________

***Conditi*ons:**

The contractor will possess the necessary insurance to protect the builder from any liability which may result from his work.

Worker's compensation will be the responsibility of the contractor.

The contractor will take all necessary precautions to protect the job site from damage and will repair or replace any property damaged through his fault.

All work will be done in strict accordance with local building regulations and the specifications in the plans.

The contractor will be responsible for cleaning up and removing all bebris from the site after work has been completed.

Other: ____

CONTRACTOR ____
DATE OF OFFER ____

SIGNATURE ____

BUILDER ____
DATE OF ACCEPTANCE ____
SIGNATURE ____

The builder reserves the right to hold ______% of the total contract price for ______ days after completion.

CONTRACT
FINISH CONCRETE FLOOR

DEADLINE DATE - TENDER MUST BE RETURNED BY________________

Plan no.__________
House Size__________

Builder
Name__________________________
Address__________________________
Phone__________________________

Contractor
Name__________________________
Address__________________________
Phone__________________________

Job location____
Lead time required____
Estimated time required to complete the job____________________________

Description of the job: (initial items included)

_____ Power or hand trowel to a smooth finish
_____ Spread, level and power tramp gravel fill to top of footing
_____ Prepare gravel for concrete floor (and build sump and cleanout boxes)
_____ Supply and convey or place gravel into basement

Additional charges

	UNIT PRICE	MEASUREMENT (YARDS-METERS)	AMOUNT (YARDS-METERS)	TOTAL
Supply Sand / Gravel	$_________	Per ___________	___________	___________
Place and Tamp	$_________	Per ___________	___________	___________
Trowel concrete floor	$_________	Per ___________	___________	___________
Other	____			

TOTAL PRICE FOR ABOVE WORK $______________

Terms of payment ________________________________
____
____

This offer will remain in effect until ________________________

***Conditions*:**

The contractor (supplier) will possess the necessary insurance to protect the builder from any liability which may result from the delivery or work.

Worker's compensation will be the responsibility of the contractor.

The contractor will take all necessary precautions to protect the job site from damage and will be responsible to repair or replace any property damaged through his fault.

All materials supplied will be in accordance with local building regulations and specifications in the plans.

The contractor will clean up and remove debris from the site at the completion of his work.

Other: ____

____

CONTRACTOR ____

DATE OF OFFER ____

SIGNATURE ____

BUILDER ____

DATE OF ACCEPTANCE ____

SIGNATURE ____

The builder reserves the right to hold ______% of the total contract price for ______ days after completion.

CONTRACT
SUPPLY BUILDING MATERIALS

DEADLINE DATE - TENDER MUST BE RETURNED BY_______________

Plan no.__________

House Size__________

Builder	***Contractor***
Name______________________________	Name____________________________
Address____________________________	Address__________________________
Phone______________________________	Phone___________________________

Job location____

Lead time required for delivery of materials ______________________________

Page 1	$ ____________
Page 2	$ ____________
Page 3	$ ____________
Page 4	$ ____________
Page 5	$ ____________
Page 6	$ ____________
Page 7	$ ____________
Page 8	$ ____________
Page 9	$ ____________
Page 10	$ ____________
Page 11	$ ____________

GRAND TOTAL FOR BUILDING MATERIALS $____________

Additional charges:

Delivery	$ ____________
Material take-off (if any)	$ ____________
Other	$ ____________
TOTAL	$______________

Terms of payment ____

____

____

*** This offer will remain in effect until ______________________________

***Conditions*:**

The supplier will possess the necessary insurance to protect the builder from any liability which may result from the delivery or work.

The supplier will take all necessary precautions to protect the job site from damage and will be responsible to repair or replace any property damaged through his fault.

All materials supplied will be in accordance with local building regulations and specifications in the plans.

Other: ____

____

SUPPLIER ____

DATE OF OFFER ____

SIGNATURE ____

BUILDER ____

DATE OF ACCEPTANCE ____

SIGNATURE ____

The builder reserves the right to hold ______% of the total contract price for ______ days after completion.

CONTRACT FRAMING

DEADLINE DATE - TENDER MUST BE RETURNED BY______________

Plan no.__________
House Size__________

Builder
Name____________________________
Address__________________________
Phone____________________________

Contractor
Name__________________________
Address________________________
Phone__________________________

Job location____
Lead time required____
Estimated time required to complete the job___________________

Description of the job: (initial items included)

Subfloor

_____ Build pony walls according to plans (if required)
_____ Construct and place wood beams, bearing walls, teleposts
_____ Install floor joists, crowns up with double joists under all parallel partitions (or install floor trusses as per design layout)
_____ Install trimmers, cross bridging, joist hangers, etc.
_____ Glue and securely nail (or screw if required) subfloor sheathing

Framing

_____ Build wall sections according to plans and specifications
_____ Place poly vapour barrier on ends of exterior walls and above partitions
_____ Place insulation where required during assembly of walls
_____ Assist in installing one piece fiberglass tub supplied by builder
_____ Install temporary bracing where required to secure partitions
_____ Complete roof framing, installation of premanufactured trusses, sheathing, ridge boards, etc.
_____ Install all windows and doors with finishing nails (countersunk)
_____ Provide backing for drywall, curtains, attic access, soap and paper holders, towel bars, closet rods, thermostat, door chimes, medicine cabinets, etc.
_____ Provide sufficient framing for soffits and fascia contractor
_____ Cut holes for cold air returns, attic vents and heat registers
_____ Install insulation stops
_____ On completion, clean job site, sweep house and stack excess lumber
_____ Frame short walls, ledge and step(s) around jacuzzi tub

Additional charges:

Build and install stairs $ ______________
Install premanufactured stairs $ ______________
Build fireplace chase with false ceiling $ ______________
Framing for dropped kitchen ceiling $ ______________
Frame exterior basement walls $ ______________
Frame furnace room $ ______________
Frame sundeck and railing (as specified on plans) $ ______________
Other (pocket doors, bow or bay windows, sunken living room, third gable, hip roof, etc.) $ ______________

TOTAL PRICE FOR FRAMING HOUSE OF SIZE
__________________SQ. FT. **$**__________

Terms of payment ______________________________

____

*** This offer will remain in effect until ____________________

***Conditions*:**

The contractor will possess the necessary insurance to protect the builder from any liability which may result from his work.

Worker's compensation will be the responsibility of the contractor.

The contractor will take all necessary precautions to protect the job site from damage and will repair or replace any property damaged through his fault.

All work will be done in strict accordance with local building regulations and the specifications in the plans.

The contractor will clean up and remove debris from the site at the completion of his work.

Other: ______________________________

CONTRACTOR ______________________________
DATE OF OFFER ______________________________

SIGNATURE ______________________________

BUILDER ______________________________
DATE OF ACCEPTANCE ______________________________

SIGNATURE ______________________________

The builder reserves the right to hold ______% of the total contract price for ______ days after completion.

CONTRACT
SUPPLY DOORS AND WINDOWS

DEADLINE DATE - TENDER MUST BE RETURNED BY________________

Plan no.___________
House Size___________

Builder
Name________________________________
Address______________________________
Phone________________________________
Job location____
Lead time required for delivery of materials ________________________________

Contractor
Name____________________________
Address__________________________
Phone___________________________

SIZE OF WINDOWS as shown on plans	SIZE quoted	DESCRIPTION	PARTICULARS	COST

Terms of payment ______________________________
____

*** This offer will remain in effect until ______________________

***Conditions*:**

The supplier will possess the necessary insurance to protect the builder from any liability which may result from the delivery or work.

The supplier will take all necessary precautions to protect the job site from damage and will be responsible to repair or replace any property damaged through his fault.

All materials supplied will be in accordance with local building regulations and specifications in the plans.

Other: ______________________________

CONTRACTOR ______________________________

DATE OF OFFER ______________________________

SIGNATURE ______________________________

BUILDER ______________________________

DATE OF ACCEPTANCE ______________________________

SIGNATURE ______________________________

The builder reserves the right to hold ______% of the total contract price for ______ days after completion.

CONTRACT
ROOFING

DEADLINE DATE - TENDER MUST BE RETURNED BY______________

Plan no.__________
House Size__________

Builder	***Contractor***
Name____________________	Name____________________
Address____________________	Address____________________
Phone____________________	Phone____________________

Job location____

Lead time required____

Estimated time required to complete the job______________________

Description of the job: (initial items included)

_____ Install asphalt shingles or other (specify) ________________

_____ Install poly in valleys

_____ Install metal flashings

_____ Install roof vents - type ______________________________

_____ Seal around all mechanical vents, stacks and chimneys

Description of materials: (to be supplied)

____

____

____

Total price for materials and labor $__________

Total price for labor only (include poly, nails,
flashing and vents but not shingles or other roofing material) $ __________

Additional charges: ____________________

____

____

Terms of payment: ____________________

____

____

*** This offer will remain in effect until ______________________

Conditions:

The contractor will possess the necessary insurance to protect the builder from any liability which may result from his work.

Worker's compensation will be the responsibility of the contractor.

The contractor will take all necessary precautions to protect the job site from damage and will repair or replace any property damaged through his fault.

All work will be done in strict accordance with local building regulations and the specifications in the plans.

The contractor will clean up and remove debris from the site at the completion of his work.

Other: ______________________________

CONTRACTOR ______________________________
DATE OF OFFER ______________________________

SIGNATURE ______________________________

BUILDER ______________________________
DATE OF ACCEPTANCE ______________________________

SIGNATURE ______________________________

The builder reserves the right to hold ______% of the total contract price for ______ days after completion.

CONTRACT
SIDING, SOFFITS, FASCIA, EAVESTROUGH

DEADLINE DATE - TENDER MUST BE RETURNED BY_______________

Plan no.__________
House Size__________

Builder
Name______________________________
Address______________________________
Phone______________________________

Contractor
Name______________________________
Address______________________________
Phone______________________________

Job location____
Lead time required____
Estimated time required to complete the job_______________________________
Description of the job: (initial items included)

Siding

_____ Install building paper behind siding
_____ Install siding as per elevations on plan
_____ Caulk all cracks around doors, windows, corners,etc.

Soffits, Fascia and Eavestrough

_____ Supply and install soffits and fascia
_____ Silicone seal all joints to intersecting roof lines and any other areas where water can enter roof from above or below shingles (or other roofing material)
_____ Supply and install eavestrough and downspouts

Description of materials: (Type - steel, aluminum, vinyl, cedar, etc.)

	Type	Grade or Brand	Color	Cost Material and Labor
Siding				
Soffits				
Fascia				
Eavestroughs				
Other				
Other				
Other				

TOTAL COST FOR MATERIAL AND LABOR $ ____________

COST FOR MATERIAL ONLY (if specified) $ ____________

TERMS OF PAYMENT ____
____
____

*** This offer will remain in effect until ______________________________

Conditions:

The contractor will possess the necessary insurance to protect the builder from any liability which may result from his work.

Worker's compensation will be the responsibility of the contractor.

The contractor will take all necessary precautions to protect the job site from damage and will repair or replace any property damaged through his fault.

All work will be done in strict accordance with local building regulations and the specifications in the plans.

The contractor will clean up and remove debris from the site at the completion of his work.

Other: ____

CONTRACTOR ____
DATE OF OFFER ______________________________________

SIGNATURE ____

BUILDER ____
DATE OF ACCEPTANCE _________________________________

SIGNATURE ____

The builder reserves the right to hold ______% of the total contract price for ______ days after completion.

CONTRACT
STUCCO AND PARGING

DEADLINE DATE - TENDER MUST BE RETURNED BY______________

Plan no.__________
House Size__________

Builder
Name______________________________
Address____________________________
Phone______________________________

Contractor
Name____________________________
Address__________________________
Phone___________________________

Job location____
Lead time required____
Estimated time required to complete the job_________________________________
Description of the job: (initial items included)

_____ Supply all scaffolding, material, and tools necessary to complete all the labor.
_____ Check framing on exterior walls is nailed solid
_____ Apply building paper from bottom up overlapping as required
_____ Firmly nail stucco wire and metal lath (if required) to all areas to be finished
_____ Apply flashing where necessary over doors and windows
_____ Supply and install scratch coat
_____ Apply trowel finish stucco (two coats) over cured scratch coat
_____ Apply parging

Description of materials:

____

____

Total price for material and labor **$** __________

Additional charges: ______________________________
____
____

Terms of payment: ______________________________
____
____

*** This offer will remain in effect until ______________________________

***Conditions*:**

The contractor will possess the necessary insurance to protect the builder from any liability which may result from his work.

Worker's compensation will be the responsibility of the contractor.

The contractor will take all necessary precautions to protect the job site from damage and will repair or replace any property damaged through his fault.

All work will be done in strict accordance with local building regulations and the specifications in the plans.

The contractor will clean up and remove debris from the site at the completion of his work.

Other: ____

CONTRACTOR ______________________________________
DATE OF OFFER ____________________________________

SIGNATURE _______________________________________

BUILDER ____
DATE OF ACCEPTANCE ______________________________

SIGNATURE _______________________________________

The builder reserves the right to hold ______% of the total contract price for ______ days after completion.

CONTRACT
MASONRY

DEADLINE DATE - TENDER MUST BE RETURNED BY_______________

Plan no.__________
House Size__________

Builder
Name______________________________
Address____________________________
Phone______________________________

Contractor
Name____________________________
Address__________________________
Phone____________________________

Job location____
Lead time required____
Estimated time required to complete the job______________________________

Description of the job: (initial items included)

_____ Supply all scaffolding, nails, ties, sand, cement, mortar
_____ Supply and install angle iron
_____ Install building paper
_____ Complete brickwork as per plan
_____ Supply and delivery of brick
_____ Other ____

Material description: (Type and amount of brick, etc.)

____

____

Additional charges: ____

____

Total price for material and labor $ ___________
Total price for labor (excluding brick) $ ___________
Unit price for brickwork $ ___________

Terms of payment: ____

*** This offer will remain in effect until ______________________________

Conditions:

The contractor will possess the necessary insurance to protect the builder from any liability which may result from his work.

Worker's compensation will be the responsibility of the contractor.

The contractor will take all necessary precautions to protect the job site from damage and will repair or replace any property damaged through his fault.

All work will be done in strict accordance with local building regulations and the specifications in the plans.

The contractor will clean up and remove debris from the site at the completion of his work.

Other: ______________________________

CONTRACTOR ______________________________

DATE OF OFFER ______________________________

SIGNATURE ______________________________

BUILDER ______________________________

DATE OF ACCEPTANCE ______________________________

SIGNATURE ______________________________

The builder reserves the right to hold ______% of the total contract price for ______ days after completion.

CONTRACT
PLUMBING

DEADLINE DATE - TENDER MUST BE RETURNED BY_______________

Plan no.__________
House Size__________

Builder
Name______________________________
Address______________________________
Phone______________________________

Contractor
Name______________________________
Address______________________________
Phone______________________________

Job location______________________________
Lead time required______________________________
Estimated time required to complete the job______________________________

Description of the job:

Supply all labor and materials to install the rough-in and all fixtures to complete the plumbing. The materials to be supplied are listed below including type of piping and brand name and description of fixtures.

Main Bathroom______________________________

____

____

____

Kitchen______________________________

____

____

____

Master Bath (ensuite)______________________________

____

____

____

Basement

____

____

Other ____________________

____

____

____

Item	Included in tender	Additional charge
Outside service hook-up	______	______
3-piece rough-in in basement	______	______
Dishwasher connection	______	______
Two lawn services	______	______
Gas connection (s)	______	______
Under floor drains	______	______
All clean outs, valves, and vent pipes	______	______
Co-ordination - plumbing & gas inspections	______	______
Travelling time to the site	______	______
Connection of gas fireplace or log lighter	______	______
Garberator hook-up	______	______
Trenching for service installation	______	______
Other	______	______

TOTAL PRICE - ALL MATERIAL AND LABOR INCLUDED $ ______

Terms of payment: ____________________

Guaranty or warranty ____________________

*** This offer will remain in effect until ____________________

***Conditi*ons:**

The contractor will possess the necessary insurance to protect the builder from any liability which may result from his work.

Worker's compensation will be the responsibility of the contractor.

The contractor will take all necessary precautions to protect the job site from damage and will repair or replace any property damaged through his fault.

All work will be done in strict accordance with local building regulations and the specifications in the plans.

The contractor will clean up and remove debris from the site at the completion of his work.

Other: ____________________

CONTRACTOR ____________________
DATE OF OFFER ____________________
SIGNATURE ____________________

BUILDER ____________________
DATE OF ACCEPTANCE ____________________
SIGNATURE ____________________

The builder reserves the right to hold ______% of the total contract price for ______ days after completion.

CONTRACT
WATER AND SEWER

DEADLINE DATE - TENDER MUST BE RETURNED BY______________

Plan no.__________
House Size__________

Builder	*Contractor*
Name______________________	Name______________________
Address______________________	Address______________________
Phone______________________	Phone______________________

Job location____

Lead time required____

Estimated time required to complete the job______________________

Description of the job: (initial items included)

_____ Trench from house to city connection

_____ Supply and connect all services, water, sanitary and storm sewer pipes

_____ Obtain necessary inspections on services

_____ Backfill all trenches

Description of materials: ______________________________________

____

____

Total price for material and labor to install
water, sewer and storm sewer services $ ______________________

Unit cost (include measurement)
for water and sanitary sewer $ ______________________

Unit cost for storm, water and sewer $ ______________________

Additional charges:

Unit price to dig water and sewer trench	$ ______________________
Unit price to dig electrical trench	$ ______________________
Winter digging	$ ______________________
Other	$ ______________________
Total additional charges	**$** ______________________

Terms of payment: ______________________________________

____

*** This offer will remain in effect until ______________________

Conditions:

The contractor will not pile fill on adjacent property without approval from the builder
The contractor will possess the necessary insurance to protect the builder from any liability which may result from his work.
Worker's compensation will be the responsibility of the contractor.
The contractor will take all necessary precautions to protect the job site from damage and will pay to repair or replace any property damaged through his fault including sidewalks and services.
All necessary permits will be obtained and all work will be done in strict accordance with local building regulations and the specifications in the plans.
The contractor will not backfill services until an approved inspection has been obtained.

Other: ______________________________
____
____

CONTRACTOR ______________________________

DATE OF OFFER ______________________________

SIGNATURE ______________________________

BUILDER ______________________________

DATE OF ACCEPTANCE ______________________________

SIGNATURE ______________________________

The builder reserves the right to hold ______% of the total contract price for ______ days after completion.

CONTRACT
HEATING

DEADLINE DATE - TENDER MUST BE RETURNED BY______________

Plan no.__________
House Size__________

Builder	***Contractor***
Name____________________	Name ____________________
Address____________________	Address ____________________
Phone____________________	Phone____________________

Job location ______________________________

Lead time required______________________________

Estimated time required to complete the job______________________

Description of the job: (initial items included)

_____ Provide builder with a heating layout
_____ Install furnace, chimneys and ductwork
_____ All joints to be screwed and taped
_____ All ductwork to be securely supported
_____ Power humidifier
_____ Dryer vent to exterior of house
_____ Venting for furnace and hot water tank
_____ Insulate ductwork where necessary
_____ Supply thermostats
_____ Storm collars where necessary
_____ Install ductwork for kitchen and bathroom exhaust fans
_____ Supply registers and grills
_____ Other

Description of furnace and materials to be supplied:

____
____
____

Additional charges:

Travel and delivery to job site	$ __________
Heat loss calculations and layout	$ __________
Power humidifier	$ __________
Thermostat type______________	$ __________
Dryer vent	$ __________
Radiant floor heating (attach specifications)	$ __________
Other	$ __________

Total price for heating **$** __________

Guaranty or warranty ______________________________

Terms of payment: ______________________________

____

*** This offer will remain in effect until ______________________________

***Conditi*ons:**

The contractor will possess the necessary insurance to protect the builder from any liability which may result from his work.

Worker's compensation will be the responsibility of the contractor.

The contractor will take all necessary precautions to protect the job site from damage and will repair or replace any property damaged through his fault.

All work will be done in strict accordance with local building regulations and the specifications in the plans.

The contractor will clean up and remove debris from the site at the completion of his work.

Other: ____

CONTRACTOR ______________________________________
DATE OF OFFER ____________________________________
SIGNATURE ____

BUILDER ____
DATE OF ACCEPTANCE ________________________________
SIGNATURE ____

The builder reserves the right to hold ______% of the total contract price for ______ days after completion.

CONTRACT
ELECTRICAL

DEADLINE DATE - TENDER MUST BE RETURNED BY________________

Plan no.__________
House Size__________

Builder
Name______________________________
Address______________________________
Phone______________________________

Contractor
Name______________________________
Address______________________________
Phone______________________________

Job location____
Lead time required____
Estimated time required to complete the job______________________________

NUMBER	ITEM	MATERIAL DETAILS
	Circuit panel	
	Light outlets	
	Switches	
	3-Way switches	
	Duplex receptacles	
	Telephone outlets	
	Cable TV outlets	
	Range service	
	Dryer service	
	Ground fault circuits in	
	Bathrooms	
	Outside	
	Jacuzzi	
	Set door chimes	
	Bathroom fans(ductwork by other)	
	Garage door opener plug	
	Smoke alarms	
	Vacuum system outlet	
	Pot lights (box framing by other)	
	Furnace	
	Dishwasher	
	Range	
	Garberator	
	Microwave oven	
	Built-in oven	
	Hood fan (ductwork by other)	

Description of the job: (describe items included)

Additional items:	***Initial items included in tender***	***Extra Charge***
Obtain electrical permits	________	$ ________
Supply & install temporary power	________	$ ________
Install poly hats around plugs	________	$ ________
Fixture hanging	________	$ ________
Supply standard fixture package	________	$ ________
Fixture allowance included	________	$ ________
Travelling to site	________	$ ________
Other	________	$ ________
Total additional charges		**$** ________

TOTAL PRICE FOR ALL ELECTRICAL **$** ________

Terms of payment: ________

Guaranty or warranty ________

*** This offer will remain in effect until ________

Conditions:

The contractor will possess the necessary insurance to protect the builder from any liability which may result from his work.

Worker's compensation will be the responsibility of the contractor.

The contractor will take all necessary precautions to protect the job site from damage and will pay to repair or replace any property damaged through his fault.

All work will be done in strict accordance with local building regulations and the specifications in the plans.

The contractor will clean up and remove debris from the site at the completion of his work.

Other: ________

CONTRACTOR ________
DATE OF OFFER ________
SIGNATURE ________

BUILDER ________
DATE OF ACCEPTANCE ________
SIGNATURE ________

The builder reserves the right to hold ______% of the total contract price for ______ days after completion.

CONTRACT
CAULKING, INSULATING & VAPOUR BARRIER

DEADLINE DATE - TENDER MUST BE RETURNED BY______________

Plan no.__________
House Size__________

Builder	***Contractor***
Name____________________________	Name__________________________
Address__________________________	Address_________________________
Phone____________________________	Phone__________________________

Job location____
Lead time required____
Estimated time required to complete the job___________________________
Description of the job: (initial items included)

_____ Caulk all areas on exterior walls between two adjacent studs, top and bottom plates
_____ Caulk and seal all holes through floor and ceiling plates
_____ Insulate all exterior walls, around windows, behind bathtubs and electrical outlets
_____ Insulate and poly fireplace chase up to false ceiling or chimney cap
_____ Install 6 mil poly, ceiling first then walls, joining over solid backing and seal with acoustical caulking
_____ Install ceiling insulation

*** **Price for labor only $** ______________

	Material Description	Amount	Cost
Insulation - Walls			
Insulation - Ceilings			
Poly walls	6ml		
Poly ceilings	6ml		
Caulking			
Other			

Total cost (materials only) $ ____________

** **Total price for material and labor** **$** __________

TERMS OF PAYMENT ______________________________________
____
____

*** This offer will remain in effect until ______________________________

***Conditi*ons:**

The contractor will possess the necessary insurance to protect the builder from any liability which may result from his work.

Worker's compensation will be the responsibility of the contractor.

The contractor will take all necessary precautions to protect the job site from damage and will pay to repair or replace any property damaged through his fault.

All work will be done in strict accordance with local building regulations and the specifications in the plans.

The contractor will clean up and remove debris from the site at the completion of his work.

Other: ____

CONTRACTOR ____
DATE OF OFFER ____

SIGNATURE ____

BUILDER ____
DATE OF ACCEPTANCE ____

SIGNATURE ____

The builder reserves the right to hold ______% of the total contract price for ______ days after completion.

CONTRACT
DRYWALL

DEADLINE DATE - TENDER MUST BE RETURNED BY______________

Plan no.__________
House Size__________

Builder
Name______________________________
Address____________________________
Phone______________________________

Contractor
Name____________________________
Address__________________________
Phone___________________________

Job location____
Lead time required____
Estimated time required to complete the job_______________________________
Description of the job: (initial items included)

_____ Supply and install 1/2" drywall to exterior walls, interior walls and ceilings
_____ Install all sheets with appropriate size drywall screws only
_____ Three coats of drywall mud or filler on all joints
_____ Sand smooth after each coat to be finished ready for paint
_____ Texture all ceilings (texture pattern___________________________)
_____ Alternate ceiling texture (room & pattern________________________)
_____ Clean up and remove debris from site

Additional charges:

Supply and install 5/8" drywall on ceilings	$ ____________
Touch-up visit on completion of finishing trade	$ ____________
Decorative ceiling	$ ____________
Delivery	$ ____________
Travelling	$ ____________
Other	$ ____________
*** **Total price for materials and labor**	**$** ____________

TERMS OF PAYMENT ____
____
____

*** This offer will remain in effect until ___________________________

Conditions:

The contractor will possess the necessary insurance to protect the builder from any liability which may result from his work.

Worker's compensation will be the responsibility of the contractor.

The contractor will take all necessary precautions to protect the job site from damage and will pay to repair or replace any property damaged through his fault.

All work will be done in strict accordance with local building regulations and the specifications in the plans.

The contractor will clean up and remove debris from the site at the completion of his work.

Other: ____

____

CONTRACTOR ____

DATE OF OFFER ____

SIGNATURE ____

BUILDER ____

DATE OF ACCEPTANCE ____

SIGNATURE ____

The builder reserves the right to hold ______% of the total contract price for ______ days after completion.

CONTRACT
CABINETS, VANITIES AND COUNTERTOPS

DEADLINE DATE - TENDER MUST BE RETURNED BY________________

Plan no.__________

House Size__________

Builder	***Contractor***
Name______________________	Name______________________
Address______________________	Address______________________
Phone______________________	Phone______________________

Job location____

Lead time required____

Estimated time required to complete the job______________________

Description of the job:

Supply and install kitchen cabinets, vanities and countertops complete with all trim and hardware as per layout supplied.

Description of materials:______________________________________

____

____

____

____

____

____

____

____

Additional charges:

Post formed countertops (or other)	$ __________
Kitchen extras (ie. spice rack / counter saver)	$ __________
Delivery	$ __________
Installation	$ __________
Other	$ __________
Total price for material and labor	**$** __________

TERMS OF PAYMENT ________________________________

____

____

*** This offer will remain in effect until ______________________

***Conditions*:**

The contractor will possess the necessary insurance to protect the builder from any liability which may result from his work.
Worker's compensation will be the responsibility of the contractor.
The contractor will take all necessary precautions to protect the job site from damage and will pay to repair or replace any property damaged through his fault.
All work will be done in strict accordance with local building regulations and the specifications in the plans.
The contractor will clean up and remove debris from the site at the completion of his work.
Other: ____

____

SUPPLIER ____

DATE OF OFFER ____

SIGNATURE ____

BUILDER ____

DATE OF ACCEPTANCE ____

SIGNATURE ____

The builder reserves the right to hold ______% of the total contract price for ______ days after completion.

CONTRACT
FLOOR COVERINGS

DEADLINE DATE - TENDER MUST BE RETURNED BY_______________

Plan no.__________
House Size__________

Builder
Name______________________________
Address______________________________
Phone______________________________

Contractor
Name______________________________
Address______________________________
Phone______________________________

Job location____
Lead time required____
Estimated time required to complete the job______________________________

Description of the job:

Supply and install undercushion, broadloom, linoleum and tiles (if contracted) with all necessary sealant, trims, and divider strips as per plan.

Description of materials:

Material (s)	Color/brand Quality	Price per sq./m.	Amount Required	Total cost
Broadloom 1				
Broadloom 2				
Broadloom 3				
Broadloom 4				
Linoleum 1				
Linoleum 2				
Linoleum 3				
Hardwood				
Tile 1				
Tile 2				
Installation:				
Broadloom				
Underpad				
Other				
Other				
Other				

Additional charges:

Premeasuring house	____________	$ ____________
Metal strips	____________	$ ____________
Stringers	____________	$ ____________
Steps	____________	$ ____________
Preparing subfloor	____________	$ ____________
Delivery and travelling	____________	$ ____________
Return visit to	____________	$ ____________
Re-stretch carpet	____________	$ ____________
Stairs to basement	____________	$ ____________
Other	____________	$ ____________
Total additional charges		**$** ____________

TERMS OF PAYMENT ____________________________________

____

*** This offer will remain in effect until ____________________________

***Conditions*:**

The contractor will possess the necessary insurance to protect the builder from any liability which may result from his work.

Worker's compensation will be the responsibility of the contractor.

The contractor will take all necessary precautions to protect the job site from damage and will pay to repair or replace any property damaged through his fault.

All work will be done in strict accordance with local building regulations and the specifications in the plans.

The contractor will clean up and remove debris from the site at the completion of his work.

The amounts of materials listed will be a sufficient amount to complete the job as per plan.

Other: ____

____

SUPPLIER ____________________________________

DATE OF OFFER ____________________________________

SIGNATURE ____________________________________

BUILDER ____________________________________

DATE OF ACCEPTANCE ____________________________________

SIGNATURE ____________________________________

The builder reserves the right to hold ______% of the total contract price for ______ days after completion.

CONTRACT
INTERIOR FINISHING

DEADLINE DATE - TENDER MUST BE RETURNED BY______________

Plan no.__________
House Size__________

Builder
Name______________________________
Address______________________________
Phone______________________________

Contractor
Name______________________________
Address______________________________
Phone______________________________

Job location____
Lead time required____
Estimated time required to complete the job______________________________

Description of the job: (initial items included)

_____ Install and case all interior doors including hardware
_____ Install and case all closet doors including hardware
_____ Install shelving, rods and hardware
_____ Install window casings
_____ Install baseboards
_____ Install all bathroom accessories
_____ Install exterior door hardware
_____ Install attic hatch and trim
_____ Install handrails
_____ Install wood railings
_____ Install underlay in kitchen, bathrooms and front entry
_____ Install all trim necessary to have a completely finished home
_____ Other ____

Additional items: (feature walls, book shelves, sunshine ceiling, etc.)

____
____
____
____
____

Total additional charges **$** __________

Description of materials to be supplied- (if contracted)

____
____
____
____
____
____

Details of staining and lacquering: (if contracted)

____
____
____
____
____

Total price for material and labor **$** __________
Total price for labor only **$** __________

Terms of payment ____
____

*** This offer will remain in effect until ______________________________

Conditions:
The contractor will possess the necessary insurance to protect the builder from any liability which may result from his work.

Worker's compensation will be the responsibility of the contractor.

The contractor will take all necessary precautions to protect the job site from damage and will repair or replace any property damaged through his fault.

All work will be done in strict accordance with local building regulations and the specifications in the plans.

The contractor will be responsible for cleaning up and removing all debris from the site after work has been completed.

Other: ____

CONTRACTOR ____
DATE OF OFFER ____
SIGNATURE ____

BUILDER ____
DATE OF ACCEPTANCE ____
SIGNATURE ____

The builder reserves the right to hold ______% of the total contract price for ______ days after completion.

13

RECORDING AND COST CONTROL

"Control" is one of the major functions of management and in the case of general contracting, controlling costs and spending is a definite must. In the construction of an average home, you can issue up to 100 cheques. Without some method of keeping track of these cheques and other financing costs, you can easily exceed your budget. With an accurate record of costs you are ready to make decisions regarding financial requirements, changes in materials, contractors, and costs.

You will also be prepared to go to your bank and discuss your financial plan (as shown in your records) and requirements. This is important if you are seeking additional financing. It only takes a few minutes every three or four days to maintain a good record system.

How much do I owe the plumber? Have I paid the first portion of the contract? When do I pay the 15% (check Government regulations) holdback? Did I go over or under on my construction materials estimate?
When did I last pay my supplier? All of these questions and many more will be easy to answer by working with simple cost control sheets and recording all costs during construction.

Recording and Cost Control Form

The following worksheets are your master cost control forms. They are to be used as an indexed record of the cost summary and loan calculation form which is broken up into 28 categories. It's a master record enabling you to maintain control of your budget from beginning to end. Control is made easy by comparing cheques written for a particular stage to the estimate used in the original cost summary and loan calculation form.

The first two columns can be completed before construction begins. From there on it's simply a matter of posting information from every financial transaction (e.g. paying invoices, reconciliation of your monthly bank statement, etc.)

RECORDING AND COST CONTROL

INDEX NO 1 – 28 from cost summary & loan calculation sheet	DATE of transaction	DESCRIPTION Subtrade/supplier Other	COMPANY NAME	PAYMENT DETAILS			COST CONTROL				BANKING INFORMATION			OTHER
				Cheque No.	Goods and Services Tax (GST)	AMOUNT PAID (Incl. GST)	Cumulative TOTAL PAID per Index No.	ESTIMATED COST from Estimating Checklist	MECHANIC LIEN HOLDBACK 15% of total PAID (if required)	DIFFERENCE OVER (or under) BUDGET	BANK DEPOSITS (Mortgage Draws or Interim Financing Deposits)	SERVICE CHARGES, INTEREST other Bank Charges	BANK BALANCE (separate House Account)	MISC. ITEMS, CASH PURCHASES, VISA PURCHASES, ETC.

COPYRIGHT © Regeneration 2000 Inc.™

RECORDING AND COST CONTROL

INDEX NO 1 - 28 from cost summary & loan calculation sheet	DATE of transaction	DESCRIPTION Subtrade/supplier Other	COMPANY NAME	PAYMENT DETAILS			COST CONTROL				BANKING INFORMATION			OTHER
				Cheque No.	Goods and Services Tax (GST)	AMOUNT PAID (Incl. GST)	Cumulative TOTAL PAID per Index No.	ESTIMATED COST from Estimating Checklist	MECHANIC LIEN HOLDBACK 15% of total PAID (if required)	DIFFERENCE OVER (or under) BUDGET	BANK DEPOSITS (Mortgage Draws or Interim Financing Deposits)	SERVICE CHARGES, INTEREST other Bank Charges	BANK BALANCE (separate House Account)	MISC. ITEMS, CASH PURCHASES, VISA PURCHASES, ETC.

COPYRIGHT © Regeneration 2000 Inc.™

RECORDING AND COST CONTROL

				PAYMENT DETAILS			COST CONTROL				BANKING INFORMATION			OTHER
INDEX NO 1 – 28 from cost summary & loan calculation sheet	DATE of transaction	DESCRIPTION Subtrade/supplier Other	COMPANY NAME	Cheque No.	Goods and Services Tax (GST)	AMOUNT PAID (Incl. GST)	Cumulative TOTAL PAID per Index No.	ESTIMATED COST from Estimating Checklist	MECHANIC LIEN HOLDBACK 15% of total PAID (if required)	DIFFERENCE OVER (or under) BUDGET	BANK DEPOSITS (Mortgage Draws or Interim Financing Deposits)	SERVICE CHARGES, INTEREST other Bank Charges	BANK BALANCE (separate House Account)	MISC. ITEMS, CASH PURCHASES, VISA PURCHASES, ETC.

COPYRIGHT © Regeneration 2000 Inc.™

RECORDING AND COST CONTROL

INDEX NO 1 – 28 from cost summary & loan calculation sheet	DATE of transaction	DESCRIPTION Subtrade/supplier Other	COMPANY NAME	PAYMENT DETAILS			COST CONTROL				BANKING INFORMATION			OTHER
				Cheque No.	Goods and Services Tax (GST)	AMOUNT PAID (Incl. GST)	Cumulative TOTAL PAID per Index No.	ESTIMATED COST from Estimating Checklist	MECHANIC LIEN HOLDBACK 15% of total PAID (if required)	DIFFERENCE OVER (or under) BUDGET	BANK DEPOSITS (Mortgage Draws or Interim Financing Deposits)	SERVICE CHARGES, INTEREST other Bank Charges	BANK BALANCE (separate House Account)	MISC. ITEMS, CASH PURCHASES, VISA PURCHASES, ETC.

COPYRIGHT © Regeneration 2000 Inc.™

14

JOB SCHEDULING

How do you measure your success as a builder? To some extent, success can be measured by the length of time it takes you to construct your home. The shorter the time, the more successful you are because you have proven yourself as a planner and organizer. Furthermore, you have reduced interest costs and, by moving in earlier, you might save on rent payments.

Anyone can build a house in six months or longer but it takes someone with appreciation for the important function of job scheduling to build a home in less than three months. Your pre-planning period can be three months, six months, one year or even two years. All the work you do in pre-planning will be tested by the efficiency of scheduling, follow-up, and co-ordination of the materials and labour throughout a reasonable three-month construction period. The more preparation and research before construction, the easier the process of co-ordinating the project.

JOB SCHEDULING TIPS

Begin by discussing your drawings with trades. Review the foundation plan with the cribber, the framing details with the carpenter and so on. There are always a few details on the drawings that either party will question. These pre-construction discussions will avoid many potential misunderstandings and difficulties during construction.

1. Review the Contract

Before you begin, review the contract to verify payment terms, price, the work to be completed, and any other details.

2. Identify Lead Times

Check each trade regarding the notice they require before beginning construction and the length of time required to complete the job. This information is valuable for job scheduling and co-ordination of a sequence of trades.

3. Be Flexible

Construction has setbacks due to poor weather, sickness, delays, etc. You must be prepared to accommodate any setbacks. A good schedule will allow for the fact that all trades have their own schedules to follow. It will allow sufficient time to complete each job. One good practice is to allow the maximum time required in your schedule (e.g. if framing takes 4 to 7 days). It's always better to gain a day than to call several trades and put them off for a day.

4. Preparation Between Trades

There are many small labour jobs that must be completed quickly between trades in order to reduce construction time. As an owner-builder, you will likely do these jobs yourself. Some of them include laying weeping tile, spreading gravel, cleaning up around the site, sweeping inside the house, caulking, insulating the walls, painting, and some finishing. The secret is to know exactly what work you will be doing in advance and to do only the work you are capable of doing within a reasonable amount of time.

5. Plan Inspections

Obtain all inspections at proper times and call ahead to make an appointment. The inspectors (CMHC, CITY GAS) may only come out your way once a week. Find out what notice they require and mark it on your construction schedule.

6. Go Local When Possible

The closer your trades and suppliers are to the job site, the more frequent visits they can make. When you have a problem, the site is not too far out of their way to make a visit with little advance notice.

7. Have a Plan or Job Schedule to Follow

Do a construction plan of the sequence of events on paper. There is far too much information to remember, and you can easily miss something important if it's not written down. Your written job schedule will allow you to schedule trades and make easy adjustments due to any construction setbacks. Some examples are shown in Section C.

8. Monitor - Maintain Good Communications

Maintain constant communication with your trades. Check their work and follow up on all inspections. A good practice is to keep a diary of all communication on your job schedule. You will have a record of the times you last talked and what was discussed. This simple procedure will assist your memory and can give you the upper hand during phone discussions.

9. Plan Your Holidays

Plan your holidays at the proper times. Don't take them all at the beginning because there is a great deal of work involved in the finishing and move-in stage. Co-ordinate your time off with your work and your construction schedule.

10. Pre-Construction Planning

Before you begin construction, you will want to review the following preparation steps to get your project started on the right foot and ensure a smooth running operation.

(a) Receive mortgage approval and verify inspections, specifications, and tests required by the mortgage company.
- Weeping tile requirements
- Sulphate soil test
- Soil bearing test
- Weeping tile and dampproofing inspection
- Framing inspection
- Plumbing, heating, and electrical inspections
- Insulation inspection
- Semi-final inspection requirements
- Final inspection requirements

(b) Pick up building permit. Is it a temporary permit allowing excavation only? Verify inspections and requirements regarding soil bearing and soil type tests.

(c) Verify development controls, if they exist. Do you require an elevation check? Inspect the property with the developer to note any damages to services or sidewalks.

(d) Arrange fire and liability insurance.

(e) Apply and receive approval of interim financing.

(f) Set up a line of credit with at least one major supplier (after approval of interim financing).

(g) Verify that the surveyor has the correct plot plan and grade slip. Check the following:
- Standard or reverse plan
- Measurement of side yards, setbacks
- Depth of excavation cuts
- Elevation to bottom of footing
- Elevation to bottom of basement windows (Note: If wood basement, allow room for header above window)
- Elevations at corner pins
- Elevations of sanitary and storm sewer invert
- Lot, block, plan number

(h) Pre-order items (such as windows, trusses, and cabinets) that require long lead times.

(i) Verify that all applications have been sent in for gas, electrical permits, etc.

(j) Install a sign on the lot to assist inspectors, trades, and suppliers in locating your property.

(k) Arrange temporary power to be set up on the site. Borrowing power from a neighbour is not as efficient as having your own outlets with breaker switches right on the site. The benefits of temporary power make it a worthwhile investment.

(l) Meet with your cribber and discuss the following:
- Plot plan, power supply
- Footing plan, walls, and rebar
- Joist plan and beams
- Fireplace detail
- Details of sunken rooms, basement entry, bearing walls, etc.
- Start date, forming materials required
- Payment detail

(m) Meet with your framer and discuss the following:
- Rough opening measurements of doors and windows
- One piece tub or shower (to be installed during framing stage) if required
- Changes on the floor plan
- Fireplace detail, kitchen drop ceiling, skylights, etc.
- Stairs, trusses, and delivery of windows
- Delivery of framing materials
- Proposed start date
- Payment details

(n) Verify that all trades have been notified of your construction plans and given approximate dates. Verify the start date for all trades up to the completion of the framing.

The Construction Critical Path

The critical path is the series of events you must undertake in order to complete your home in the shortest possible time. The word critical applies to functions that cannot be performed until something previous has been completed first. The items that are not critical branch off different steps of the critical path and can be completed at any time after those steps.

As can be seen in the construction critical path diagram, it is possible to have more than one activity in progress at one time. Knowledge of the order of events is necessary to build the home quickly and save on interest costs. For example, once the house is framed in, you could have the plumbing and heating contractors working on the inside while the siding and roofing contractors are working on the exterior. You need a good temporary power supply to schedule two or more trades at one time. Study the construction critical path and note the critical and the non-critical steps.

Within the construction critical path, there are many small jobs that must be completed either by a hired labourer or the owner-builder. Some of these jobs are critical steps because they precede the work of a major trade.

JOB TO BE DONE:	**PRIOR TO**
Loosen dirt under cribbed footing where water and sewer drains will eventually go	Pouring concrete footings
Tell the electrician the location of the electrical panel, Placing weeping tile and gravel, clean up around site	Before backfill
Place backing and blocking for curtains, medicine cabinets, bathroom accessories	Installing poly on exterior walls
Prepare basement floor- Gravel, build clean-out boxes, sump box if required, lay poly over gravel	Pouring concrete floor

JOB TO BE DONE:	PRIOR TO:
Install underlay (K3 board)	Installation of Kitchen Cabinets
Clean house	Painting
Prime walls	Interior finishing
Install linoleum	Installing toilets

* Doing these small items quickly will enable other trades to continue with little delay.

Creating Your Job Schedule

Right at this moment, producing a job schedule on paper may sound complicated and frightening. No need to be alarmed; this chapter provides a sample construction schedule and worksheets which will be easy for you to complete prior to construction.

The first step is to collect all your tenders and record on each the job duration and the required lead times. If you are unsure which trades you will actually use, record the maximum number of days you think should be adequate to complete the job. Refer to the Bar Charts in the scheduling construction chapter for assistance.

The second step is to apply this information to your construction critical path. You now have a total number of construction days to complete the home. Take note that these are construction days. You do not, at this time, take into consideration weekends or holidays. By using construction days your schedule remains flexible. A setback on a Monday would put you one day behind on your schedule. Don't be misled.

You must also use a calendar to schedule starting dates and deliveries. The purpose of the worksheets is to develop a workable schedule independent of construction setbacks. Once completed, you can apply start dates and completion dates (in pencil) according to your calendar.

The third step involves listing the job functions on the job schedule worksheets provided here. This list of information will be similar to the sample job schedule shown.

The fourth step is to fill in the required lead times on your schedule. For example, the drywaller is scheduled to start on day 24 and requires two weeks' notice. Back up on the worksheets and make a note to call on day 13 or 14. The two weeks' notice represents 10 working days on your job schedule worksheets. Do not count the weekends even though some trades work during these days.

The final step is to plan, with a calendar, the start dates of construction. Fill in on your worksheets the start dates up to the framing stage. As construction proceeds, you will continue to fill in dates, always keeping two to three weeks ahead of actual construction.

Bar Chart

There are many alternatives to forming a job schedule, and every builder's schedule will vary to some degree. Devise your own schedule to conform with the products and construction methods of your custom home. A bar chart is one alternative means of scheduling (see the example of a bar chart in this chapter). It clearly identifies the critical path showing which functions can be performed on the earliest possible construction days and the length of the job. The major disadvantage of the bar chart is that it leaves no room for diary information and recording many details, such as lead times.

Construction Master Plan

The construction master plan, also included in this chapter, is for you to keep track of all the construction activities including inspections, start and completion dates and a diary for making notes.

Communication is Vital to Job Scheduling

Construction Critical Path

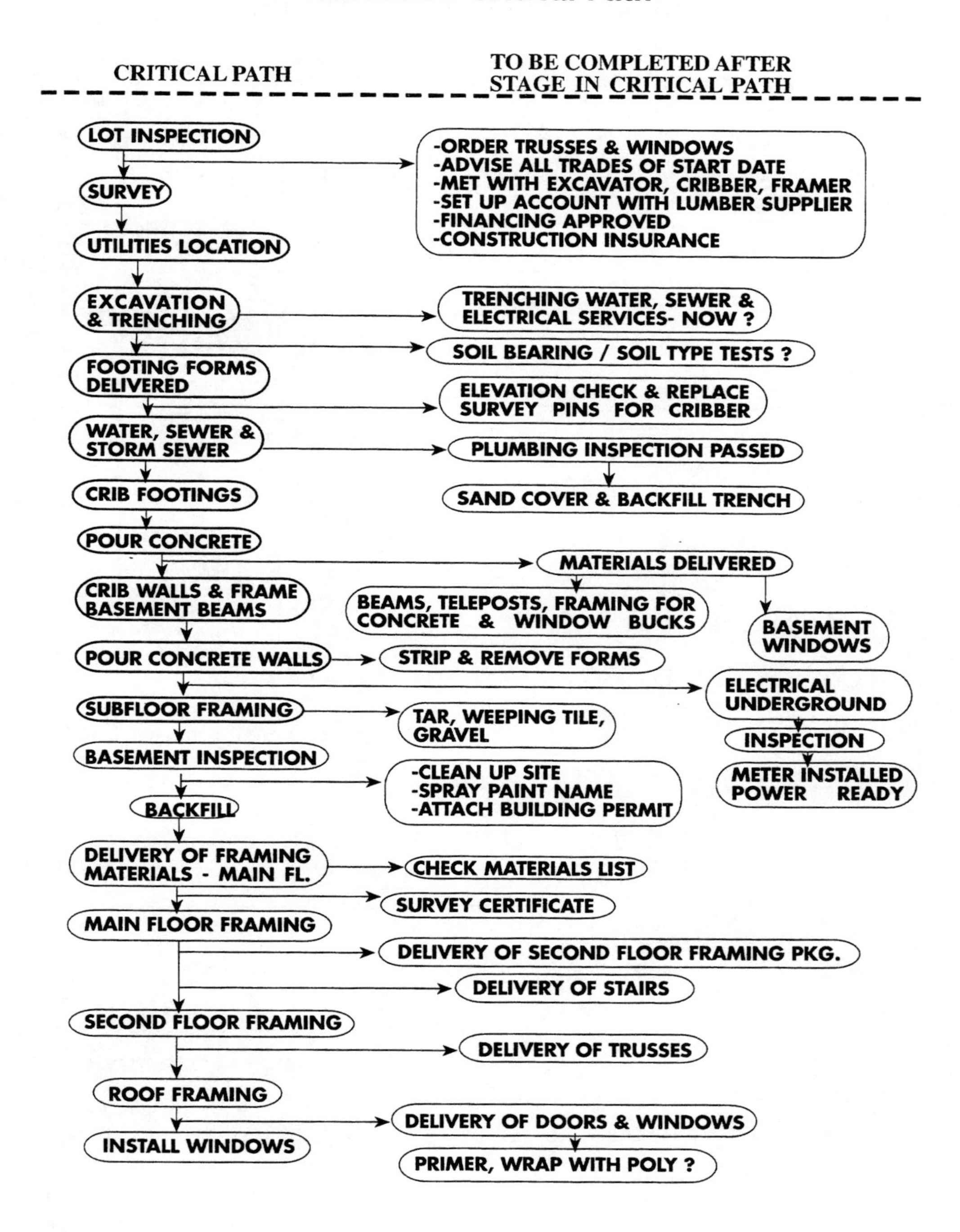

Cronstruction Critical Path (Continued)

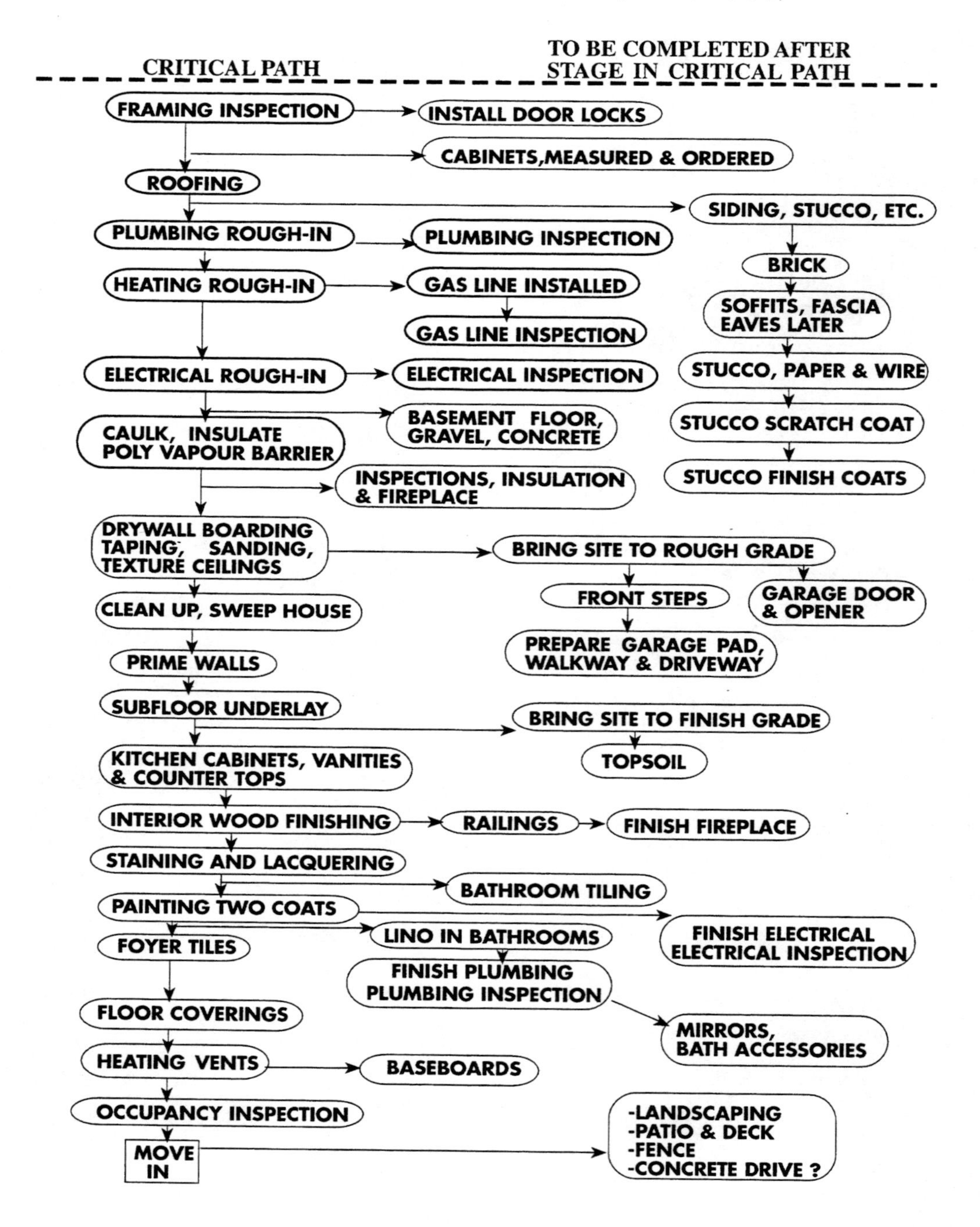

LIST OF INSPECTIONS

TYPE OF INSPECTION	WHEN	WHY	WHO CALLS FOR IT	WHO DOES THE INSPECTION
Site Inspection	Before excavation	Notify developer of any damages to services or sidewalks	Builder	Builder
Soil test and Compaction test	After excavation (only if required by Land Developer)	Maybe building on a fill site (i.e. softer ground). Soil type may require additives in concrete.	Builder (as advised by Land Developer)	Soil Engineer
Elevation Check	After excavation and before pouring footings	To ensure correct elevation before starting foundation	Builder	Surveyor
Plumbing, Water and Sewer System	After installed and before trench is backfilled	Required by Plumbing Inspection Branch to ensure proper installation and codes are met	Plumber	Plumbing Inspector
Electrical under-ground service to foundation	Before backfill of found-ation and electrical trench	Required by Electrical Inspection Branch to ensure proper instal-lation and electrical codes are met	Electrician	Electrical Inspector
Foundation Inspection	Before backfill Note: cannot backfill a preserved wood base-ment until basement floor is in place	Check workmanship on weeping tile (if required) dampproofing (tar) and gravel placement. Check quality of concrete.	Builder	City Inspector Note: may require Engineer inspection if preserved wood foundation (PWF)
Survey Certificate	After foundation is in, usually after backfilling site	To receive Survey Certificate certifying house is placed properly on lot. Required by Lawyer to advance funds from MortgageCo.	Builder	Surveyor
Framing Inspection	Before insulating	Check framing codes are correct to ensure structure is strong and safe. Check framing is done as per plans submitted. NOTE: builder must check for straightness of walls and installation of backing, blocking for drywall, etc.	Builder	City Building Inspector
Financial #1	Usually after building is closed in Whenever you request funds be advanced	To receive draw on mortgage or interim financing advance	Builder	Appraiser or Mortgage Co. Representative
Plumbing Rough-in	Before insulating	Check plumbing codes are being met	Plumber	Plumbing Inspector
Gas Inspection	After gas line hooked up and pressurized (30 Lbs air)	Check for leaks in gas line and proper installation	Plumber or	Gas Inspector from City
Electrical Rough-in	Before insulating	Check electrical codes are being met	Electrician	Electrical Inspector
Fireplace Inspection	Before boarding or covering with drywall	Check for proper installation of firestop, clearance of framing from chimney, builder has followed manufacturers directions	Builder	City Building Inspector
Insulation, caulking and vapour barrier	Before boarding	Check for holes in vapour barrier	Builder	Builder (sometimes City Bldg. Inspector
Financial #2	After cabinets installed and house has first coat of interior paint	To receive additional funding	Builder	Appraiser or Mortgage Co. Rep.
Final Electrical	After all electrical complete	Check codes, work all done as per codes, close file	Electrician	Electrical Inspector
Final Occupancy	Before moving in	Structure complete, safe, codes met	Builder	City Inspector
Final Financial # 3	When 95% Complete	Receive Remaining mortgage funds	Builder	Mortgage Appraiser

BAR CHART

Job	Construction Schedule																												Dates
Construction Days	1	2	3	4	5	6	7	8	9	10	11	12	13	14	15	16	17	18	19	20	21	22	23	24	25	26	27	28	
Survey	x																												
Excavation		x	x																										
Water & Sewer				x																									
Crib & Concrete Footings				x	x																								
Walls - Crib & Concrete						x	x	x	x	x																			
Tar, Weeping Tile, Gravel											x	x																	
Framing Subfloor											x	x	x																
Electrical Underground											x																		
Backfill & Bring to Grade													x	x															
Framing														x	x	x	x	x	x	x									
Windows & Doors																				x									
Roofing																				x	x	x	x						
Natural Gas Service																							x	x					
Soffits & Fascia																							x	x					
Masonry Work - Brick																							x	x	x				
Siding or Stucco																									x	x	x	x	
Plumbing Rough - In																					x	x	x						
Heating Rough - In																					x	x	x						
Electrical Rough - In																								x	x	x			
Basement - Gravel																										x			
Vacuum System																										x			
Intercom, Telephone, T.V.																										x			
Concrete Basement Floor																											x	x	

BAR CHART Pg 2

Job	Construction Days																												Dates
Construction Days	28	29	30	31	32	33	34	35	36	37	38	39	40	41	42	43	44	45	46	47	48	49	50	51	52	53	54	55	
Insulation, Poly	x	x	x	x																									
Garage Pad Prepare & Concrete				x	x	x																							
Prepare Front Step, Walk & Drive					x	x	x																						
Drywall, Board, Tape, Sand					x	x	x	x	x	x	x	x																	
Ceiling Insulation - Blown In									x																				
Ceilings Textured												x	x																
Primer Coat of Paint													x	x															
Kitchen Cabinets													x	x	x														
Interior Finishing, Railings, Doors														x	x	x	x	x											
Underlay															x														
Ceramic Tiles, Foyer Tiles														x	x	x	x												
Stain, Lacquer, Wood Finishing																		x	x										
Painting Interior																				x	x	x	x						
Painting Exterior																				x	x								
Linoleum (before setting toilets)																						x							
Finish Electrical / Light Fixtures																								x					
Finish Plumbing																							x	x					
Outside, Finish Grading & Loam																					x	x							
Floor Coverings, Carpet, Lino																									x	x			
Install Baseboards																										x			
Finish Heating - Registers & Vents																											x		
Touch - Up / Clean - Up																											x		
Completion Date																												x	

CONSTRUCTION MASTER PLAN

WORK REQUIRED	Site Inspection Survey Call Utility Companies Before Digging	Excavation Footing Material delivered 2 x 6 forms ?	Crib Footings Elevation & Soil Test Insp Pour Concrete Footings	Basement Wall Package Crib Walls Elect. meter support cribbed in wall 14' 2x10 P.W.F.	Pour Concrete (use Concrete Pump) Strip Forms	Water & Sewer Trenching & Service Electrical Trench dug	Subfloor Material Framing Subfloor Electrical Service & Meter Installed	Plumbing Insp. Elect. Insp. Backfill Insp. Backfill to Rough Grade	Windows Ordered Trusses Ordered Lumber Pkg Delivered	Framing Walls
DATE STARTED										
DATE COMPLETED										
INSPECTION CHECKED										
NOTES										

CONSTRUCTION MASTER PLAN

WORK REQUIRED	Premanufactured Stairs Ordered Delivery Tub Enclosure Jacuzzi	Trusses Delivered Roof Framing Package Delivered Framing Second Story & Roof	Install Doors & Windows Fireplace Installed & Framed in	Framing Deck Framing Inspection Roofing Started	Plumbing Rough-in & Inspection Heating Rough-in	Kitchen Site Measured Electrical Rough-in & Inspection	Soffit & Fascia Eaves Later	Front Steps ? Masonry or Brick Trim	Basement Gravel Concrete Basement Floor	Vacuum System, Telephone, Cable T.V., Intercom, Security System All Roughed in
DATE STARTED										
DATE COMPLETED										
INSPECTION CHECKED										
NOTES										

CONSTRUCTION MASTER PLAN

WORK REQUIRED	Stucco Paper and Wire Siding or Other	Caulking Insulation & Vapour Barrier	Drywall Board, Tape, Mud	Prepare Garage Pad Walkway and Driveway	Texture Ceilings Ceiling Insulation Blown in	Sand Walls Primer Coat of Paint	Kitchen Cabinets Delivery of Interior Finishing Materials	Finishing Carpenter Underlay, Doors, Railings, Casings, Shelving Etc.	Ceramic Tiles, Foyer Tiles, Marble Vanity Tops	Stain, Lacquer, Wood Finishing
DATE STARTED										
DATE COMPLETED										
INSPECTION CHECKED										
NOTES										

CONSTRUCTION MASTER PLAN

WORK REQUIRED	Painting Interior Painting Exterior	Linoleum (before setting toilets) Hardwood Flooring	Finish Electrical, Light Fixtures, Inspection	Finish Plumbing, Inspection	Finish Outside Grade & Loam	Floor Coverings Carpet	Mirror Doors, Mirrors Bath Accessories	Finish Heating Registers & Vents	Install Baseboards Appliances	Touch up Clean up Final Inspection Completion Date
DATE STARTED										
DATE COMPLETED										
INSPECTION CHECKED										
NOTES										

SAMPLE JOB SCHEDULE

DAY	ITEM CHECK	JOB	STARTING DATE	COMPLETION DATE	COMMUNICATION DIARY
5 Days Prior to Construction		-Receive mortgage approval -Notify all beginning trades of start date -Pick up building permit (applied for one to two weeks ago) -Set up insurance for construction and fire -Arrange interim financing -Begin to arrange and pre-order materials for delivery on a date that will be confirmed later			
4 Days Prior to Construction		-Establish line of credit with at least one major lumber supplier and one concrete company -Order windows, obtain copy of rough opening measurements for cribbers and framers -Lot inspection for damage to services and sidewalks prior to starting -Contact survey company for day 2 or day 1 -Contact excavator for day 1 and cribber for day 2 -Contact plumber for day 2 to install water and sewer service			
3 Days Prior to Construction		-Contact electrician for electrical permit to set up temporary power (best to set up your own power A.S.A.P.) -Call gas company (and other utilities) for location prior to digging -Backhoe for exposing cable to splice for power service -Apply at gas company for new service -Order basement windows for cribber			
2 Days Prior to Construction		-Contact lumber yard to deliver footing material, ladder mat, subfloor (wood basement, if applicable) for day 1 -Survey lot just prior to excavation -Contact framers -Arrange for gravel delivery -Elevation check ordered for afternoon day 1 (if required) -If soil bearing certificate is required, arrange for late afternoon day 1			

SAMPLE JOB SCHEDULE

DAY	ITEM CHECK	JOB	STARTING DATE	COMPLETION DATE	COMMUNICATION DIARY
1 Day Prior to Construction		-Approval of interim financing -Confirm excavation for day 1 concrete for footings for day 2 -If winter, coal and straw to burn ground for footings -Soil test or concrete report on type used may be required by city or - CMHC - check			
Day 1		-Start in morning -Excavation - should try to pile dirt away from water and sewer and electrical trenches -Delivery of footing material, rebar and subfloor -Basement windows for cribbers -Elevation check by engineer - late afternoon if possible or after footings have been formed -If wood basement, arrange gravel to be delivered and spread			
Day 2		-Crib footings (if wood basement, begin framing wall) -Pipe under footings to save digging by plumber -Site inspection by lender (if required) -Order gravel for inside footings for day 4 to be spread by bobcat or self -Pour concrete footings and pads - owner to specify proper type of concrete			
Day 3		-Allow footings to cure one day prior to spreading gravel -Gravel can be conveyed into house afterwards; however, this usually is more expensive -Notify framer to start day 8			
Day 4		-Crib basement walls - two days - assume not cast-in-place subfloor (if joist cast in concrete allow two additional days to crib) -Order concrete, pump, ensure proper strength and additives, e.g., sulphate resistant, winter heat, etc.			

SAMPLE JOB SCHEDULE

DAY	ITEM CHECK	JOB	STARTING DATE	COMPLETION DATE	COMMUNICATION DIARY
Day 5		-Crib walls -Order framing package -Check when trusses will be made -Confirm delivery of subfloor package			
Day 6		-Pour concrete walls. Allow one day to set -Receive subfloor basement package - check materials to packing slip			
Day 7		-Strip forms -Trenching for water and sewer when possible - have water and sewer lines placed into basement - Inspect and backfill trench - Co-ordinate timing with plumber -Confirm framer - start subfloor on day 8 -Cribbers to clear forms away from site			
Day 8		-Bearing walls, beams, subfloor -Confirm delivery of framing package -Ensure home is identified with a sign -Building permit must be displayed on property			
Day 9		-Complete subfloor -Break off snap ties -Waterproofing (or poly dampproofing for wood basement) -Weeping tile -Washed gravel over weeping tile -Call for foundation inspection prior to backfill -Try to backfill before delivery of framing materials; however, you can backfill at any time during the framing stage or when framing is complete. You should not hold up construction or framers while waiting to be backfilled.			

SAMPLE JOB SCHEDULE

DAY	ITEM CHECK	JOB	STARTING DATE	COMPLETION DATE	COMMUNICATION DIARY
Day 10		-Begin framing upper floor - exterior walls -Contact mechanical trades, roofer, siding - inform them of progress to-date -Order fiberglass tub to be delivered prior to finishing the framing (if required) -Confirm delivery of trusses			
Day 11		-Confirm delivery of doors, windows -Start framing exterior walls -Bring survey certificate, proof of insurance to lawyer and sign mortgage documents (Must be done prior to receiving mortgage draws.) -Survey certificate to building permit department (if required)			
Day 12		-Receive trusses -Framing of interior walls -Contact shingler and order shingles to be delivered and placed on the roof for day 15			
Day 13		-Framing of interior walls, trusses -Notify drywaller of progress to-date			
Day 14		-Framing - trusses, roof -Confirm plumbing, heating and electrical -Contact bank - insure no problems with interim account -Receive windows and doors, installed by framing crew -Gas line to house - mark exact location on outside wall - co-ordinate with plumber or gas fitter - Gas line from furnace to outside wall, installation of meter, test line, gas inspection			
Day 15		-Roofing -Rough-in plumbing -Call for order and measure of front concrete step (if pre-cast) -Order insulation for day 19 -Rough-in heating; in winter hang from joists, in summer wait until basement floor is poured			

SAMPLE JOB SCHEDULE

DAY	ITEM CHECK	JOB	STARTING DATE	COMPLETION DATE	COMMUNICATION DIARY
Day 16		-Complete roofing -Complete rough-in of plumbing and heating -Order interior gas line -Blocking, backing, caulking -Check with lender to ensure mortgage advances are being made			
Day 17		-Rough-in electrical -Deliver and install front concrete step -Interior blocking, backing, caulking			
Day 18		-Complete rough-in electrical, call for inspection (if required) -Clean up -Confirm delivery of insulation -Order cabinets, pre-measure -Installation of fireplace			
Day 19		-Begin siding, soffit, fascia - usually after backfill -Inspection - town or city prior to insulating -Prepare for concrete floor - level gravel, clean out and make sump boxes -Lay poly over gravel bed before pouring concrete floor			
Day 20		-Siding, brick, stucco - mark location of each product -Insulating and poly -Confirm concrete for floor - day 21 - advise other trades who will be working in the house, frost must be out of ground, gravel level, services inspected, and teleposts straight -In winter, interior gas line should be in, furnace hung, gas meter installed ready to light furnace as soon as insulation is in place or prior to pouring concrete floor			
Day 21		-Siding -Insulating, complete poly -Clean up -Pour concrete floor			

SAMPLE JOB SCHEDULE

DAY	ITEM CHECK	JOB	STARTING DATE	COMPLETION DATE	COMMUNICATION DIARY
Day 22		-Inspection from your lender on insulating and poly -Siding -Contact gas company - may require temporary heat for drywall -Electrical hook-up - splice for power -Delivery of drywall confirmed			
Day 23		-Siding, eaves -Install door locks on front -Electrical inspection -Begin drywall - boarding			
Day 24		-Boarding -Contact interior finisher -Front steps (formed, wood, pre-cast concrete, fiberglass, brick)			
Day 25		-Drywall - tape roughcoat -Arrange for delivery of all interior doors, casing, baseboards, shelving, hardware, etc.			
Day 26		-Drywall - tape rough coat -Grading of lot - hauling in fill or hauling away fill			
Day 27		-Drywall - sand, tape second coat -Owner can be staining interior finishing material in basement of house - ready to install (if method applies, check other alternatives of interior finishing)			
Day 28		-Drywall - sand second coat -Confirm delivery of cabinets for day 32 or 33			
Day 29		-Drywall tape - final third coat -Purchase material for painting			
Day 30		-Drywall - sand final coat -Texture ceilings -Arrange for telephone - new number and send out change of address cards			

SAMPLE JOB SCHEDULE

DAY	ITEM CHECK	JOB	STARTING DATE	COMPLETION DATE	COMMUNICATION DIARY
Day 31		-Clean up - vacuum drywall dust -Prime walls - roll or spray and then sand -Order loam for lot- level			
Day 32		-Paint - do yourself -Deliver cabinets to house			
Day 33		-Paint			
Day 34		-Paint -Contact electrician re: completion date			
Day 35		-Installation of telephone -Interior finishing -Underlay K3 board under linoleum			
Day 36		-Pre-measure for carpets -Cabinets - begin installation, plumber delivers fixtures, templates -Mirror - owner measure, order and install			
Day 37		-Cabinets, countertops, templates cut for plumber -Inspection 70% lender -Interior doors, casings, bi-folds -Door locks, stops, other hardware -Attic hatch -Window casings -Arrange delivery of dishwasher and range hood prior to plumbing and electrical finals			
Day 38		-Finish plumbing fixtures - lino in first, then toilet -Baseboards -Closet shelves, rods -Railings			

SAMPLE JOB SCHEDULE

DAY	ITEM CHECK	JOB	STARTING DATE	COMPLETION DATE	COMMUNICATION DIARY
Day 39		-Handrails -Complete all interior woodwork -Confirm installation of carpets - day 43 -Electrical fixtures, complete			
Day 40		-Complete all interior woodwork -Complete exterior grading, sidewalk, parking pad or driveway -Ensure you have had final inspection from: electrical, plumbing, heating, and all other authorities having jurisdiction			
Day 41		-Complete all interior woodwork and clean up -Prepare subfloor for carpets			
Day 42		-Interior ceramic tiles -Towel bars, medicine cabinets, paper roller			
Day 43		-Delivery of carpets, install nailer strips			
Day 44		-Carpet installation -Lino installed before toilet -Install heat vents			
Day 45		-Drywall touch up -Finish installation of carpets -Complete above			
Day 46		-Receive 100% inspection from mortgage company, town or city (if required)			
Day 47		-Move in			
Day 48 Day 49		-Complete any deficiencies -Follow up on payments, mechanic's lien holdback, advances, legal documents, interest adjustment and first mortgage payment date, change of address, etc. -Begin landscaping, fence, driveway, garage - when affordable			

Job Scheduling Sequence - Pre-construction Events

- Plans and plot plan or site plan
- Plans approved - (required if building in an architecturally controlled sub-division)
- Receive grade slip (not applicable on acreages)
- Building permit application approved
- Financing approved - mortgage & interim financing, Lawyer named
- Set up seperate chequing account
- Line of credit with lumber supplier, concrete, windows, cabinets
- Confirm contracts with most trades & suppliers
- Confirm WCB coverage
- Arrange construction and liability insurance
- Plan start date, check lead times for trades
- Notify trades of start date, order windows, trusses, cabinets
- Temporary power required? (Cribbers may supply a generator for basement only?) - Have installed in basement right away.
- Call (Utility Companies) before you dig for service locations
- Site Inspection, sidewalks, curbs, services (return of any damage deposit paid to Land Co.) - Watch for cracked sidewalks, etc.
- Surveyor requires plans, site plan and grade slip
- Meet with excavator to walk site and discuss job, trenching now or later - Black Loam brought in + pushed to back for yard.
- Applcation for utilities, power, gas, water, phone, cable - Could take 2 months+.

Construction - To Subfloor Stage

- Survey 1 day prior to excavation
- Excavation (and trenching for services now or later ?)
- Delivery of lumber and rebar for cribbing footings (use 2 x 6 boards) - Concrete walls, include 1-2x8 14' (pressure treated) for electric meter
- Deliver basement windows or basement window frames or include 2x8 lumber to build window bucks on site
- Elevation check by surveyor just at start of the cribbing.
- Soil inspection by builder (engineer) if poor conditions exist (i.e. soft sand, loam)
- Soil bearing certificate required? Compaction test required?
- Crib footing (cribber) advise cribber of weeping tile requirements to prepare footings
- Water and sewer lines under footings now or later (depends on trenching)
- Water and sewer lines inspected (called by Plumber)
- Water and sewer trench backfilled after inspection
- Pour concrete footing
- Strip footing forms, 2x6 material will be used up during construction
- Crib concrete walls
- Pour concrete walls, strip & remove forms usually the next day
- Delivery of main subfloor (beams, floor joists and plywood)
- Framers begin to install subfloor
- Electrician installs outside service to pressure treated 14' 2x8 and orders meter from city
- Builder marks tar line with spray paint
- Install rigid insulation for frost protection if required for walk out basement
- Tab snap tie holes and spray tar or use foundation wrap
- Weeping tile installed (if required)
- Clean gravel (1" washed rock) over weeping tile
- Electrical service inspected (inspection called by Electrician)
- Foundation inspection by city prior to backfill (called by Builder)
- Post building permit, mark name on foundation
- Backfill and rough grade site (if wood basement do not backfill until after floor is poured)
- Contact surveyor for survey certificate, provide copy to lawyer with proof of construction and house insurance, sign mortgage documents at Lawyer's office
- Financial inspection to obtain first mortgage advance (called by Builder)

Need Survey + Insurance

Job Scheduling Sequence - Framing to Lock-up

- Lumber package delivered (main floor walls)
- Framing walls and garage
- Tentative delivery date set for roof trusses and windows (ordered earlier)
- Delivery of second floor framing materials, floor trusses, sheathing, beams, etc.
- Delivery of roof trusses and delivery of roof framing materials
- Pre-manufactured stairs ordered and installed by Carpenters Brydon Stairs Co.
- Delivery and installation of large tub enclosure, shower enclosure (protect with insulation & cardboard)
- Meet Framers on site (coffee) to discuss how fireplace, drop ceiling, tub surround and other items should be framed including interior decorative items such as indirect lighting
- Delivery and installation of fireplace, frame around face and hearth
- Framing second storey and roof
- Delivery of windows and doors to be installed by Carpenters, order garage door (remove screens)
- Install temporary locks, (supplied by finishing materials supplier) obtain 5 identical keys
- Measure and order metal fireplace chase top, if required
- Scrap in garbage bin on site, save longer pieces for blocking and backing

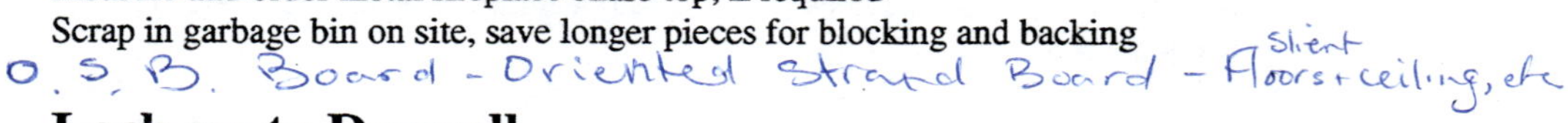

Lock-up to Drywall

- Plumbing rough-in (note: some overlap, plumbing & heating starts before Framers leave)
- Heating rough-in (furnace is hung from ceiling), interior gas line installed
- Framers finish odds and ends, drop ceiling, jet tub surround, blocking , sundeck, exterior battons, stucco relief, brick mould, garage door trim, and other interior and exterior items
- Drywaller visits home for a board count, checks backing and other framing
- Finisher visits home to check door sizes, Hardwood installer measures & orders hardwood
- Delivery and installation of Garage door (door opener installed later, after drywall)
- Electrician and Builder (owner) walk through to identify locations (lights, plugs, switches)
- Kitchen measured for pre-manufactured cabinets
- Inspections for plumbing and interior gas line connection (called by Plumber)
- Electrical rough-in, plugs, switches, lights, furnace, fireplace fan, etc.
- Rough-in Vacuum system (plan ahead for concrete floor, drywall, interior finishing)
- Rough-in Telephone
- Rough-in Cable T.V.
- Rough-in Intercom system
- Rough-in Security and Home theatre systems
- Electrical inspection (called by Electrician)
- Framing inspection (called by Builder), Framers return to fix any deficiencies
- Financial inspection by Bank Representative or Appraiser to receive second mortgage advance (called by Builder)
- Exterior Soffits and Fascia, Eavestrough to be installed after exterior is finished
- Concrete front steps supplied and installed, formed for concrete or built by Carpenters
- Brick or other masonry work
- Clean interior to prepare for basement gravel and drywall
- Basement gravel, poly, prepare (wood boxes for cleanout access) pour and finish concrete floor (if temporary heat required, pour floor after insulation)
- Interior Caulking, Poly vapour barrier, Insulation
- Exterior siding or Stucco paper and wire or other exterior finish
- Outside Gas-line installed by Gas Company, heating contractor returns to ignite furnace
- Drywall board, tape , mud and sand - three coats
- Outside preparation of garage slab, walkway and driveway
- Textured ceilings, choose style, consideration to interior design
- Ceiling / attic insulation blown in

Interior Finishing - Painting

- Primer coat of paint
- Delivery & installation of hardwood flooring, previously stored in warehouse, moisture level checked
  have in house at least 2 weeks before installation to see how it will react to climate.
- Delivery of interior finishing materials
- Interior finishing carpenters install underlay in kitchens, bathrooms, foyer and any other areas requiring tile, marble or linoleum
- Kitchen cabinets, bathroom and laundry room vanities delivered and installed
- Kitchen and bathroom countertops installed by cabinet company, marble or other supplier
- Railings and handrails installed by finishing carpenters or other supplier
- Install interior doors, door and window casings, interior closet shelving, book shelving, baseboards installed later (if white laquered, install baseboards now for a finished look)
- Kitchen cabinet installer cuts out sink templates in arborite countertops, templates left by plumber
- Other interior finishing, fireplace mantel, ceiling trim, etc.
- Bathroom Ceramic tiles, foyer tiles, kitchen backsplash tiles, kitchen floor tiles all supplied by tile supplier and installed by a qualified tilesetter
- Clean up, Drywallers return to touch up any nicks in the walls prior to final coat of paint
- Stain, lacquer wood finishing, interior doors, casings, bookshelves, baseboards, etc.
- Outside finish coat of stucco and parging
- Concrete driveway and walks prepared and finished, block off driveway for min. 7 days to allow concrete to cure and harden
- Eavestrough installed after all other exterior siding, brick and trim
- Painting interior walls
- Financial inspection for mortgage draw

Painting - Move In

- Linoleum before setting toilets
- Delivery of electrical light fixtures (supplied by builder / owner)
- Delivery of built-in dishwasher to be installed by plumber and connected by electrician
- Finish electrical, plugs, switches, lights, panel
- Finish plumbing, toilets, sinks, taps, hot water heater, dishwasher, garborator (all fixtures and taps supplied by plumber)
- Finish vacuum , intercom, security, telephone, cable T.V. systems
- Plumbing and electrical inspections
- Outside - painting front door, garage door, window and other trims
- Measure, supply and install mirrors, by-pass mirror closed doors, other mirrors
- Finishing carpenter installs hardware, door locks, bathroom accessories, closet shelving and rods and new exterior door locks to replace the temporary construction locks
- Clean home thoroughly
- Finish applied to hardwood floor (no other trades allowed in the home during finishing)
- Outside - finish grading and loam
- Carpet supplied by retailer and installed by floor covering mechanic
- Finish heating, clean ducts and install heat registers and cold air returns
- Install baseboards not installed earlier
- Touch up, clean house
- Final inspection by city inspector
- Financial inspection for mortgage advance
- Install appliances

Move in

- Final mortgage advance after Lien holdback period (example 10% holdback for 45 days)
- Other items - landscaping, sod, trees, fence, and interior blinds and curtains, decorating

15

CONSTRUCTION TIPS

Use the checklists on the following pages to make sure that your construction is progressing smoothly and in the best way possible. Good luck!

EXCAVATION / BACKFILL / GRADING CHECKLIST

- Check the type of machine best for your job. It may be possible to have a backhoe excavate and dig trenches all at the same time. In winter you may require a cat to dig through frost. Discuss how the machine will enter onto your property without damaging services or sidewalks.
- Make certain excavation will not be too low or too high, check cuts to present existing homes
- Check to see that fill is not piled on adjacent lots.
- Discuss jogs in walls, garages, services, fireplace chase, etc.
- Check placement of fill to allow ease of backfill.
- Discuss the finished grade for drainage.
- Locate access ramps, place for materials to be dropped off on site and location of service trenches before digging.
- Check for damage to sidewalks or services before you start and after. Take pictures, notify developer or whoever holds a damage deposit.
- Check for sufficient space between foundation and sides of excavation for cribbers to work in.
- Check for hauling away fill or hauling in extra fill.
- Obtain an elevation check before pouring concrete footings.
- Check building requirements for a soil bearing test prior to placement of footings.
- Before backfill, check to ensure electrical services are in.
- Check to see if you will require window wells.
- Check grading after completion - should be smooth. The better the grading job, the better the loam coverage.

CRIBBING CHECKLIST

- Verify dimensions of footings with cribber, check detail of fireplace, garage, telepost pads, bearing walls. If winter, protect from frost.
- Check for temporary power for cribbers.
- Check for footing material, sill plates, ladder material, and rebar on site for cribbers.
- Check placement of basement windows and door bucks.
- Check that forms are oiled and in good shape to prevent the loss of concrete.
- Check for proper placement of rebar, especially around corners and openings.
- Check that all footing post pads are at correct locations.
- Check to ensure joists are level and in the right places (assuming cast-in-place subfloor, all crowns in joist lumber should face upwards).
- Check for tub drain, chimney space, no joists in the way, all double joists in place.
- Advise cribber to clean away all waste material from in or around the basement.
- Examine poured basement wall after stripping forms and advise cribbers or concrete company to repair all voids and honeycomb areas.
- Allow concrete seven days to cure before backfill.

DAMPROOFING / WEEPING TILE / GRAVEL CHECKLIST

- Ensure snap tie holes are individually sealed before applying two coats of dampproofing to entire wall from footing up to **grade** level. Mark wall to show height of coating.
- Check that footing wall junction is well sealed.
- Check local weeping tile requirements. Are there storm sewers?
- Check weeping tile: Should be straight and start at a high point on opposite side of house from storm sewer connection sloping toward the connection or sump box.
- Check coverage of gravel over weeping tile for 6 inches of washed gravel.
- Check alternative methods of placing gravel within basement. May be best to spread after footings and before cribbing walls. This will reduce shovelling by hand.
- Check the level of the gravel bed according to height of drain, height of furnace and stairs (if installed).
- Tamp or vibrate gravel smooth and hard then re-check the level of the gravel to ensure even spread of concrete.
- Six mil poly placed between gravel and basement floor.
- Check for plumbing inspection **before** pouring concrete floor.

SLAB ON GRADE
(Houses Without Basements)

- Take every precaution to ensure dryness.
- Must have a well-drained site with the finished landscape grade sloping away from the house for adequate surface drainage.
- Install a base course of clean gravel or crushed rock. The gravel base will absorb water

and keep the slab dry.

- Install a drainage system to carry away any excess water.
- Install a 6 mil polyethylene vapour barrier just prior to pouring the slab to prevent moisture entering the slab from the soil below.
- Check the soil condition.
- There are many methods of constructing a slab on grade foundation depending on soil conditions, requirements for frost coverage, requirements for drainage, and the intended use of the building, e.g. summer camp or year-round home. Research books containing technical data on slab on grade foundations and consult your designer and an engineer. The engineer can provide useful data on soil compaction and soil types and make recommendations for better construction. It's always best to seek professional advice before spending a large amount of money and possibly making costly construction mistakes.

FRAMING CHECKLIST

- Meet with framer to review start date, when materials will arrive on site. Go over plans re: exterior walls, partitions, stairs, beam detail, trusses, sizes of interior doors, drop ceilings, etc.
- Advise framer if there is a fiberglass tub or shower.
- Give framer a copy of **the rough opening measurements** for windows and doors; windows to arrive on site on last day of framing.
- Check placement of poly around perimeter of subfloor, under exterior wall plates, behind end studs of interior partitions against outside walls, and on top plates of inner partitions.
- Check for poly under subfloor on all heated cantilevers and overhangs.
- Check backing for drywall, blocking for curtain rods, closet rods, medicine cabinets, towel bar, paper holder, attic hatch, thermostat, door chimes, drop ceilings, and soffits and fascia on the exterior.
- Check for false ceiling in fireplace chase at ceiling height; discuss with framer.
- Check nailing of subfloor, remove nails missed, check for spots requiring additional nailing.
- Check framing method of eaves or ladders. Is it strong enough?
- Check for ridge blocks between trusses to support trusses and roof sheathing.
- Ensure trusses are secured to gable ends and trusses are toe-nailed to inner partitions.
- Check for installation of insulation stops.
- Check for cut-outs for cold air returns and heat registers.
- Check walls for bends due to warped studs.
- Check level of door openings and squareness of bathroom wall where mirror will be installed.
- Install joist hangers where floor joists are not supported.
- Check all nailing of sill plates, bottom plates to subfloor around door openings, beams, top plates, etc.
- Wall for main plumbing stack must be 2 x 6 or larger to accommodate width.
- Check that floor joists are lined up or laid out so that they are not under the waste outlet of your toilet(s).
- Check that there are no floor joists parallel and under the plumbing wall.

ROOFING CHECKLIST

- Ensure the framing is capable of supporting the type of roofing material intended. A cedar shake roof will require 1/2" roof sheathing versus the common 3/8" sheathing. A tile roof will require closer spacing of roof trusses to support the extra weight.
- Use a heavy shingle #235 lb. on low slope roofs.
- Roofs with 4/12 slope or under should have 6 mil polyethylene eaves protection.
- Check for proper colour co-ordination of roofing material before ordering.
- When owner-builder is supplying shingles, have them delivered and hoisted up onto the roof.
- A heavy interlocking shingle will have fewer blow-offs and make for a better, longer lasting roof.
- Check for placement of poly and tar in all valleys.
- Ensure all flashings, collars, and seals are securely installed.

SIDING CHECKLIST

- Ask siding contractor when to order materials.
- Ensure your colours conform to the area and your architectural controls.
- Ask your siding contractor if the following should be installed before siding:
  - Hydro stack and meter box
  - All outside electrical outlets (unless marked clearly on the outside)
  - Dryer vents, fresh air intake, etc.
  - Outside water taps
  - All windows and doors complete with brick mould
  - Lot is backfilled (depending on grade levels)
- Mark location of brick or other types of siding on walls with spray paint.
- Check for installation of building paper before siding.
- Check for all necessary pieces of trim on corners, around windows and doors, starting strips, etc.
- Check material after installation for cracks or dents and notify contractor to repair.
- Check all openings around trim work, doors, and windows. Ensure caulked and sealed on the edges to prevent the passage of water.

  ***Important:** Caulk all cracks, joints, holes around fascia joints & roof corners.
- Inform contractor to remove all garbage and debris at completion of work.
- Check for damages after grading is complete.

PLUMBING CHECKLIST

- Check location of water and sewer connections on property and calculate the distance from inverts to house.
- Co-ordinate with plumber the trenching and installation of services under the footing.
- Check, if winter, if ground is frozen. If so then dirt under footing, where services will go, should be loosened prior to pouring concrete. This will ease digging afterwards.
- Check for weeping tile requirements and whether or not your property has storm

sewers.

- Plan and co-ordinate with plumber exact location of basement rough-in, floor drain, and washing machine drain.
- Check for required plumbing inspections prior to covering sewer lines in the basement and prior to backfill water and sewer trench.
- See framing checklist; check to ensure all walls and floor joists are lined up properly to accommodate plumbing stack and toilet waste outlet. Any large bathtubs, jacuzzi or one-piece showers are installed during the framing stage.
- Mark location of outside water services and gas meter on outside of house. Check gas meter location with gas fitter and gas company.

HEATING CHECKLIST

- Don't just buy any furnace. There are many new models on the market which are high efficiency and energy saving. Carefully investigate several systems and consult with more than one heating contractor.
- The installation of a gas system will require a separate permit. Check with your plumber or gas fitter.
- For your protection; the installation must be done by a certified gas fitter who can assume the responsibility for ensuring that the gas fitting is completed in a safe manner in accordance with the applicable safety regulations.
- Consult your heating contractor regarding furnace capacity, size and location of flue, location of furnace room, combustion air and fresh air intakes, cold air returns, heat registers, dryer vent and extras (e.g. gas log lighter, power humidifier, air-to-air heat exchanger).
- Check that size of furnace is adequate but not oversized for a well sealed and insulated home.
- Mark location of gas meter on outside wall. Check regulations regarding location.
- In winter, hang furnace from floor joists; in summer, pour concrete floor first.
- After installation of furnace, check for damage.
- After home is complete, check furnace filter. It may need replacement due to a heavy accumulation of dust during construction.
- Afterwards, replace any bridging or blocking removed by heating contractor.

ELECTRICAL CHECKLIST

- Discuss electrical layout and location of switches with draftsman. Your tender for electrical will be based on the information provided in the plans.
- Prior to start, check with the electrical inspection department (ask at your municipal office) for their rules and regulations on underground services, temporary power, inspections, and any code requirements.
- Consult with electrician regarding plans, inspections, and temporary power.
- Check the location of the thermostat; should be centrally located on an interior wall on the main floor and away from any direct sunlight.
- Check to ensure sufficient number of circuits in the kitchen to accommodate appliances.
- Check for ground fault system for bathroom plug for razor.

- Rough-in extra items at the beginning (e.g.. ceiling fan, dishwasher, freezer outlet) to save expense.
- Notify framers of any dropped fluorescent ceilings or pot-lights to ensure framing preparations for electrician.
- Always set up temporary power first to guarantee a continuous, reliable source for the trades. It is an inexpensive service for providing trades with sufficient outlets and power from a neighbour.
- Mark location of electrical panel on basement plan, convenient for future development.
- Mark location of exterior lights and plugs.
- Instruct your mechanical trades that holes drilled in roof, floor or ceiling members (trusses, floor joists) are not to be larger than 1/4 depth of the member and located not less than 2 inches from the edge.
- Check for underground inspection prior to backfill of electrical trench.
- After rough-in check all outlets, switches, and plugs are in accordance with your print. Check wire location for outside plugs, dryer, stove, microwave, furnace, range hood, garberator, dishwasher, fridge, smoke detector, door bells, chimes.
- Check outlets for television and telephone. Contact proper department for any prewire services that you may require.

CAULKING / INSULATING / POLY CHECKLIST

Closing off avenues of escape of warm air is a high priority in cutting heating costs. In the average home, nearly every hour, or more than 20 times a day, the entire volume of air inside the house is replaced by air from outside.

The energy efficient features may add $1 to $2 extra per square foot, but reducing air changes can cut energy bills for heating from $800 a year to less than $200. Furthermore, you have the added benefits of an increased comfort level, a quieter home, and an increased potential for re-sale.

The requirements for a low energy home are:

(a) An efficient design
(b) Efficient windows located on the south side
(c) Thicker insulation
- Walls minimum R20 insulation
- Ceilings R40 (use high heel trusses to maintain R value over exterior walls)

(d) Increased air tightness
- 6 mil poly vapour barrier for complete protection
- Well caulked and sealed exteriors

(e) More efficient and smaller heating equipment and water heaters
(f) Use of an air to air heat exchanger for controlled ventilation and heat recovery from exhaust air

- Prior to installing windows, caulk exterior, wrap with poly (18-inch roll), staple to window. After installing window, insulate around the jamb on the inside, fold poly back and staple to framing members. This method prevents any drafts from the outside.

- Attic hatch should be framed up 18" with plywood to prevent blown-in insulation from falling down hatch.
- Caulk all areas on exterior walls between two adjacent studs, top wall plates, and bottom plate.
- Caulk all holes through floor and ceiling plates to seal as a fire stop.
- Check for care taken in installing insulation.
- Insulate and drywall fireplace chase up to chimney cap.
- Check insulation around doors and windows between tar paper and stud, not between tar paper and window frame. Do not insulate tightly around doors and windows or they will bow out of shape.
- Check insulation blown in ceiling after drywall for proper thickness.
- Insulate and poly basement walls.
- Check for poly behind all electrical outlets in outside walls and ceilings.
- Check for proper installation of poly - ceiling first, then walls (use a heavy grade, e.g.. 6 mil)
- Check poly for holes before drywall.

DRYWALL CHECKLIST

- Check straightness of walls prior to boarding by placing a long straight edge or 2 x 4 perpendicular to wall studs.
- Check for completion of all framing including blocking, backing, boxed in ceilings, fireplace, etc.
- Check how long the job will take to complete.
- Sweep house clean before delivery of material.
- Mark on subfloor the height of blocking for future reference during interior finishing.
- Do you require heat? If necessary, co-ordinate temporary heat with taper.
- Should have heat on a couple of days before boarding to dry out the wall studs.
- Avoid using propane if possible; it releases moisture.
- Check tender for materials supplied - 5/8 instead of 1/2 inch board is better on ceiling for strength and prevents sagging.
- Request drywall contractor to take an accurate board count for you if you are supplying the materials.
- Have board placed into rooms, large sheets into the largest rooms.
- Inform drywall boarder not to puncture poly during installation.
- Do not cut away poly from around window openings; it provides protection for windows while home is being painted.
- Inform the contractor of ceilings not to be textured.
- Check required corner beads on closet openings.
- Check nailing and screwing of board (nail first around edges, then securely fasten with rows of drywall screws).
- Check walls and corners for smoothness.
- Check sanding after completion.
- Co-ordinate clean-up and removal of garbage.
- In the meantime, do what other jobs you can do on the exterior, e.g. grading, steps, painting trims and windows, etc.

INTERIOR FINISHING CHECKLIST

- There are numerous types of interior finishing products. Research several alternatives and select the one most suitable and affordable.
- Check codes for proper height of handrails, height of shelves, and maximum distance between spindles.
- Discuss with your finisher the most appropriate finishing method.
  - Prime walls, install unstained finishing material, stain and lacquer finishing, paint walls (in that order)
  - Prime and paint walls, paint and stain finishing material separately, install finished material, touch-up nail holes, etc. (in that order)
  - Prime and paint walls, stain finishing material separately, install, touch-up nail holes and varnish (in that order)
  - Prime and paint walls, install prefinished interior finishing material, touch-up nail holes (in that order)
- Allow sufficient time to complete all the finishing. Once you move in, the unfinished jobs never seem to get completed.
- The finished look is very important. Learn the techniques of finishing from your carpenter rather than from your own trial and error mistakes. If you have no experience, you can easily ruin expensive products.

FLOOR COVERINGS CHECKLIST

- Plan location of seams to be in less noticeable places.
- Get a guaranteed price! Salesperson's estimate may differ from installer's.
- Installer should pre-measure home before cutting carpets. Exact measurements may vary slightly from measurements on blueprints.
- Check to ensure contract includes all extras, e.g. nailing strips, stringers, steps, underpad, and installation.
- A good quality underpad (3/8 or 1/2 inch rubber) will provide better cushion and extend carpet life.
- Check subfloor nailing and smoothness.
- Sweep and vacuum before delivery of floor coverings.
- Tiled floors must have a strong plywood underlay with no movement.
- In bathrooms to be carpeted, toilet must be installed first. In bathrooms using lino, the lino is installed first.
- The more neutral the colour of carpets, the more acceptable for resale. Colours of floor coverings should compliment tiles, drapes, woodwork, and furniture.
- Obtain a guarantee on the installation. Carpet may require re-stretching later.

16

CONTRACTING WITH A BUILDER OR PROJECT MANAGER

Armed with good plans, specifications and all the helpful advice presented in this management system, you are also now in a position to contract with a builder or hire a project manager.

Good contracts make good friends !

Included in this chapter are two contracts, the first one is to be used for a complete custom home where the builder takes control of the whole project and delivers to you a "turn key" complete home. The second contract is for hiring a project manager who will consult you throughout the construction. It is different from the first contract in that you are the builder and you will pay all the subtrades directly. The project management agreement allows you to have more flexibility and control during construction.

CONSTRUCTION CONTRACT

THIS AGREEMENT made in duplicate this _____ day of ____________, 19___.

BETWEEN:

____

Name of Contractor (hereinafter called the "Contractor")

OF THE FIRST PART

- and -

____

Names of Purchaser (hereinafter called the "Purchasers")

OF THE SECOND PART

WHEREAS the Purchasers are entitled to become the registered owners of the following described lands:

____

Building Address

in the City / Township of ______________________________ , in the Province / State of _________________ (hereinafter called "the said lands"); and

WHEREAS the Contractor has agreed with the Purchasers to construct a dwelling on the said lands in accordance with the terms and conditions hereinafter set forth:

NOW, THEREFORE, THIS AGREEMENT WITNESSED as follows:

ARTICLE 1 - PLANS AND SPECIFICATIONS

1.01 The Contractor agrees to build a dwelling house on the said lands in accordance with the detailed working plans and specifications approved by the Mortgagee named in the Mortgage registered, or to be registered against the said lands. Each of the parties hereto acknowledge as having received the said plans signed by the other party. Specifications are attached hereto and marked as **Schedule "A"**.

1.02 The price to be paid to the Contractor hereunder shall be the sum of ___________ ____DOLLARS.

Written amount in dollars

ARTICLE 2 - TERMS OF PAYMENT

2. The Purchasers covenant to pay the said price together with interest at the rate of TWO (2%) PER CENT per annum above the prime lending rate of interest then in effect at the ______________________ (*Name of Bank, Branch, City, Province or State*) on any portion thereof which is not paid when due pursuant to this Agreement, payment to be made in the manner prescribed in **Schedule "B"** attached hereto.

ARTICLE 3 - MORTGAGE APPROVAL, CANCELLATION, PERMIT, SPECIAL CONDITIONS

3.01 IT IS EXPRESSLY AGREED AND UNDERSTOOD that this Agreement is conditional upon the Purchasers being approved as Mortgagors of the said lands on or before ________, ________, ________ and that such Mortgage or Mortgages will be in an amount of not less than ______________________________ DOLLARS. In the event of such approval not being granted within the time limited, or in the event of the Purchasers failing to execute the required mortgage documents forthwith after having been approved, at the option of the Contractor this agreement shall be null and void and upon the Contractor having paid to the Purchasers such sums as they shall have paid to the Contractor pursuant to the terms hereof, both parties shall be absolutely released and discharged from all further obligations hereunder. In the event of the Purchasers obtaining a 'completion mortgage' the PURCHASERS HEREIN SPECIFICALLY ACKNOWLEDGE that they are and will be responsible for arranging interim financing.

3.02 The Contractor has provided allowances for the following:

Light Fixtures	$__________	Other $	___________
Appliances	$__________	Other $	___________
Landscaping	$__________		

ARTICLE 4 - MORTGAGE PROCEEDS

4.01 The Purchasers covenant forthwith upon receipt thereof to assign to the Contractor all proceeds of the said Mortgage to which the Contractor is entitled pursuant to the terms of this Agreement and further do covenant to assign the proceeds payable under such fire insurance policies respecting said dwelling as may be required to rebuild the said dwelling house.

ARTICLE 5 - UTILITIES, FEES, INTEREST AND INSURANCE

5.01 A. The Contractor Agrees To Pay:

(i) All utility connections, but not so as to restrict the generality of the foregoing, electricity, natural gas, water, sanitary and storm sewers, where applicable but not including telephone connections;

(ii) Liability insurance covering the said lands and dwelling house, until occupancy by the Purchasers, but the Contractor shall be liable to the Purchasers only for such loss as may be covered by such policy and assumes no responsibility whatsoever for any loss or damage arising out of the construction of the said dwelling house other than as herein before provided;

(iv) All fire insurance premiums respecting said lands from the date hereof until occupancy by the Purchasers;

(v) All legal fees and disbursements with respect to progress advances as referred to in Schedule "B" hereto.

B. In Addition To The Said Price, The Purchasers Shall Pay:

(i) All taxes, rates and assessments pertaining to the said lands during the construction period;

(ii) Any other solicitors' fees and disbursements covering preparation and registration of documents respecting the said lands, other than those fees and disbursements as outlined in **Article 5.01** (a) (v) above;

(iii) Any Mortgage insurance premiums deducted from the Mortgage proceeds pursuant to the provisions of the National Housing Act.

(iv) All interest, if any, charged by the Mortgagee in connection with the Mortgage to build the said dwelling house on the said lands.

ARTICLE 6 - COMPLETION AND DELAY

6.01 The Contractor agrees to build the said dwelling house in strict accordance with the detailed plans and specifications attached as Schedule "A" to this agreement, in a proper and workmanlike manner and with all due diligence and dispatch and to have the said dwelling house completed *One Hundred Twenty (120) days* after commencement of construction, provided the Contractor does not guarantee the completion of the dwelling house or the possession thereof to the Purchasers if the reasons for the delays in construction are caused by unfavourable weather, strikes, fires, shortages of material, acts of God or any other causes whatsoever beyond the control of the Contractor. The Contractor agrees to pay all reasonable expenses incurred by the Purchasers if the delay in construction and completion of the dwelling house is attributable to and within the control of the Contractor. The Contractor assumes no liability of responsibility whatsoever for any materials or services supplied at the request of the Purchasers without the written consent of the Contractor first being obtained.

6.02 In the event that the Contractor becomes aware that the dwelling house cannot be completed on or before *One Hundred Twenty (120) days* after commencement of construction, the Contractor shall immediately notify the Purchasers of (a) the reason for the delay, and (b) the later date on which the Contractor anticipates the dwelling house can then be completed.

6.03 The Contractor shall keep his work areas in tidy condition and free from the accumulation of waste products and debris, and upon completion of the work, the

Contractor shall remove from the site all tools, construction equipment, and any waste products or debris and leave the work area in a clean and tidy condition.

7.01 Any variations, additions, corrections or removals from the aforesaid plans and specifications shall be effected only by an amending agreement in writing signed by the parties hereto and shall, if required, be subject to acceptance by the said Mortgagee. The full estimated cost of any such variations, additions, corrections or removals as aforesaid shall be paid by the Purchasers to the Contractor prior to the commencement by the Contractor of the work required to effect any such variations, additions, corrections or removals.

7.02 In the event that such variations, additions, corrections or removals are not completed at the time or times contemplated by the provisions of Article 2 of this Agreement for the payment of monies by the Purchasers to the Contractor, the value of such variations, additions, corrections or removals not yet completed shall also be deducted from the monies to be paid by the provisions of Article 2 of this Agreement. Should there be any variance between the estimated and actual cost of such work, same shall be adjusted between the parties hereto accordingly. In the event of any dispute as to any such variance, such dispute shall be settled by arbitration in the manner provided in Article 9.02 of this Agreement.

ARTICLE 8 - SURVEY

8.01 The location of the dwelling house on the said lands, as well as any other improvements to the said lands, shall be in accordance with the plot plan for which any development or building permit has been issued by the proper Governmental Authority. The Contractor shall at the Purchasers' request provide the Purchasers with a survey certificate confirming the necessary setbacks and sideyards.

ARTICLE 9 - ACCESS, PREPOSSESSION, OCCUPANCY, INSPECTION AND ARBITRATION

9.01 The Purchasers covenant and agree to give free uninterrupted and exclusive possession of the said lands from the date hereof until the Contractor has completed the dwelling house in accordance with the terms hereof; provided, however, that the Purchasers shall have the right of inspection at all reasonable times, so long as the Purchasers do not interfere with or interrupt the work of the Contractor. Upon completion of the said dwelling house and its final inspection by the said Mortgagee's inspector and provided the Purchasers are not in default hereunder, the Contractor shall surrender exclusive possession of the dwelling house to the Purchasers. Prior to the surrender of such possession, the Contractor shall, at the request of the Purchasers, join with the Purchasers in a final inspection of the said lands and dwelling at the conclusion of which the Purchasers shall be entitled to submit in writing to the Contractor, such complaints, objections and deficiencies, if any, as they may have observed with respect to the materials and construction thereof. Any valid complaints or objections so stated shall be corrected or remedied by the Contractor as soon as practical thereafter.

9.02 Any disputes to the validity of any such complaint or objection, or as to the parties' respective responsibilities concerning same shall be settled by arbitration in the manner following: Either party may serve upon the other a written notice requiring the matter to be arbitrated and such notice shall set out the name of the arbitrator appointed by the party giving such notice; within Seven (7) days of the receipt of such notice the other party shall appoint its arbitrator and notify the first party of the name of same. The two persons so appointed shall select a third arbitrator and shall thereupon proceed with all due diligence to hear and settle the dispute. A decision of the majority of the arbitrators shall be final and binding upon the parties and the provisions of the Arbitration Act of state or province shall apply in all respects save as herein before varied.

9.03 The Contractor shall be utterly absolved from any liability or responsibility whatsoever with respect to any complaints or objections not referred to on the statement mentioned in paragraph 9.01 above respective of the time at which such complaints or objections are made, except for any latent defects in materials and construction as should subsequently come to the Purchasers' attention.

9.04 In the event the Purchasers shall occupy the said dwelling without having requested the inspection referred to in the above three preceding paragraphs, the Purchasers shall be deemed to have accepted the said dwelling in the state in which they find it and shall be deemed to have waived all rights to object to or complain about any defect in workmanship or construction as if such inspection had not occurred.

ARTICLE 10 - WARRANTIES

10.01 Insofar as manufacturers' warranties permit, the Contractor shall assign the same to the Purchasers provided, however, it shall not be responsible for performing or undertaking any manufacturer's warranty or guarantee for any machinery or equipment installed in the said dwelling house.

ARTICLE 11 - DEFAULT BY CONTRACTOR

11.01 If the Contractor should be adjudged bankrupt or make a general assignment for the benefit of his creditors, or if a receiver should be appointed, or if the Contractor should neglect to prosecute the work in accordance with the terms hereof, or fail to make prompt payment to sub-contractors, material men or labourers, the Purchasers may serve upon the Contractor a written notice requiring him to cure the default or neglect specified in such notice within seven (7) days of the delivery thereof and if the Contractor should fail to comply with the valid requirements of such a notice within the time so limited, this Agreement shall forthwith be terminated and the Purchasers may take possession of the premises and of all materials, tools and appliances thereof and finish the work in accordance with the said plans and specifications in such manner as they may deem expedient, but without undue delay or expense. In such event, the Contractor shall not be entitled to any further payment hereunder, but upon completion of the work, an accounting shall be had between the Purchasers and the Contractor in which the costs of completion necessarily incurred by the Purchasers shall be set off against the balance due the Contractor at the time the Purchasers took possession in accordance with this paragraph. If the unpaid balance of the contract price shall exceed the expense of finishing the work, such excess shall be paid to the Contractor; if such expense shall exceed such unpaid balance, the Contractor shall pay the difference to the Purchasers.

ARTICLE 12 - DEFAULT BY PURCHASERS

12.01 In the event that Purchasers should default in any of the covenants or agreements herein, the Contractor may at its option cease work and treat the contract as repudiated forthwith on the occurrence on such default, and the Contractor may recover payment for the work already completed proportionally to the total contract price plus damages, including loss of profit together with interest thereon that may be due and payable under the provisions of this Agreement or as reasonable for delay in payment, such interest to be at the rate of TWO (2%) per cent per annum above the Prime Lending Rate of interest then in effect at the ______________ *name of branch ,*
______________ *location , city , province/state .*

ARTICLE 13 - PURCHASERS' REPRESENTATION

13.01 Until they shall have paid the said price due hereunder in full, the Purchasers shall not:

(a) cause any of the proceeds of the said Mortgage to be held back by said Mortgagee;

(b) sell the said lands or premises or enter into any agreement for sale whereby the title may be encumbered by a caveat;

(c) encumber the said lands in any other manner.

13.02 The Purchasers hereby authorize, direct and empower the Contractor to do all such acts and complete all such forms as may be necessary to have the premises inspected by the Mortgagee from time to time as advances are made under the terms of the Mortgage.

ARTICLE 14 - QUALITY OF WORKMANSHIP

14.01 The Contractor guarantees that the workmanship used in the erection of the dwelling house shall be as good as or better than the housing standards of the Division of National Research and the Contractor will provide the Purchasers with a warranty having the same standards and conditions as that provided under the New Home Certification Program of ______________ (Province / State) but in any event, the Contractor shall be responsible for faulty materials and workmanship for a period of One (1) year and for a further period of Five (5) years for structural defects. The Contractor further guarantees that the building materials shall be of the standard provided by the specifications attached to this Agreement as Schedule "A" or of a standard equivalent to CMHC Housing Specifications, whichever is higher.

ARTICLE 15 - ENTIRE AGREEMENT

15.01 This Agreement shall constitute the entire contract between the parties hereto and no representation, warranties or statements made by any employee or agent of the Contractor other than those in writing signed by the Contractor shall be binding on the Contractor so as to vary the terms hereof.

ARTICLE 16 - NO ASSIGNMENT

16.01 Neither party to this Agreement shall assign the same without written consent of the other, provided however, that nothing herein contained shall be construed so as to restrict the right of the Contractor to employ sub-contractors in the construction of the said dwelling house.

ARTICLE 17 - NOTICE

17.01 Any notice required to be given pursuant to the terms hereof shall be given by either party hereto in writing and mailed or delivered to the other at the following addresses:

Contractor's Address: ______________________________

____

____

Purchaser's Address: ______________________________

____

____

Any notice so delivered by mail shall be deemed to have been received by the other party forty-eight (48) hours after same has been posted in a prepaid envelope addressed as aforesaid.

ARTICLE 18 - BINDING EFFECT

18.01 This Agreement shall extend to, be binding upon and enure to the benefit of the heirs, executors, administrators, successors and assigns of the parties hereto.

ARTICLE 19 - INTERPRETATION

19.01 All words in this agreement may be read and construed in singular number instead of the plural if there be less than two Purchasers named and in such case, this agreement shall be deemed to bind the Purchasers individually as well as severally and jointly and also the masculine gender shall be construed to include the feminine or a body corporate where the context of this agreement so requires.

ARTICLE 20 - TIME ESSENCE

20.01 It is agreed that time is to be considered of the essence of this agreement.

IN WITNESS WHEREOF the parties hereto have hereunto executed these presents all on the day and in the year first above-written.

NAME OF BUILDING CONTRACTOR

PER: ______________________________

[illegible]

SIGNED, SEALED AND DELIVERED in the presence of:	)	
	)	
______________________________	)	______________________________
Witness	)	Purchaser
	)	
______________________________	)	______________________________
Witness	)	Purchaser

SCHEDULE "B"
SCHEDULE OF PROGRESS PAYMENTS
(Example based on $120,000 construction costs)

1. DEPOSIT - Preparation, Permits — AMOUNT $ 1,000.00

2. Basement ready for Backfill Est.% 16 — AMOUNT $ 19,040.00
Less 15% Holdback $ 2,856.00
Net Advance $ 16,184.00

Excavation ______
Foundation ______
Waterproofing ______
Weeping Tile ______
Subfloor ______

3. Interior ready for Drywall Est.% 25 — AMOUNT $ 29,750.00
Less 15% Holdback $ 4,462.50
Net Advance $ 25,287.50

Backfill ______
Framing, Sheathing ______
Roof completed ______
Rough-in Plumbing ______
Rough-in Heating ______
Rough-in Electrical ______
Insulation, Vapour Barrier ______
Exterior Doors, Windows ______
Basement Floor Poured ______

4. Intermediate Inspection State Est.% 29 — AMOUNT $ 34,510.00
Less 15% Holdback $ 5,176.50
Net Advance $ 29,333.50

Drywall Completed ______
Interior Walls, Ceiling
Finished ______

5. House Complete - Ready For Occupancy Est.% 30 — AMOUNT $ 35,700.00
Less 15% Holdback $ 5,355.00
Net Advance $ 30,345.00

Interior Doors Hung ______
Floors Finished ______
Electrical Completed
(including fixtures) ______
Plumbing completed
(including fixtures) ______
Kitchen Cupboards Installed______
Exterior Completed ______
Grading Completed ______
Sidewalks ______
Drywall Completed ______
Heating Completed ______
All Trim Completed ______

6. RELEASE OF HOLDBACK 35 DAYS AFTER COMPLETION ______ — AMOUNT $ 17,850.00

TOTAL $120,000.00

PROJECT MANAGEMENT AGREEMENT

This agreement made the _______ day of ___________________, _______.

Between:

____

(Name of Project Manager)

(hereinafter referred to as the Project Manager/Consultant)

OF THE FIRST PART

- AND -

____

(Owner)

____

(Owner)

(hereinafter referred to as the "Owners")

OF THE SECOND PART

WHEREAS the owners have agreed to construct a residential dwelling. AND whereas the owners have agreed to engage the Project Manager to carry out the services detailed in this Agreement which are agreed to by both parties.

CHECKLIST OF PROJECT MANAGER'S DUTIES AND SERVICES

ARTICLE 1 CONSULTANT'S DUTIES AND SERVICES

___ ___ **A)** Provide consulting on construction methods and products to provide working specifications

___ ___ **B)** Oversee and coordinate those suppliers and trades (contracted with the owners) and to report to the owners on a timely basis as to a scheduled progress of the Project. Ensuring the coordination of the various Contractors and Subcontractors the Project's timely completion and conformance with the construction schedules and that such work is carried out in conformance with all applicable laws, statutes, ordinances, regulations, and by-laws of the federal, provincial and municipal authorities and with any lawful direction of any public offices. Report to the Owners in the event that any party retained by the Owners is not providing the services he has contracted to provide on a timely basis.

___ ___ **C)** Provide advice in land planning and negotiation land purchase.

___ ___ **D)** Coordinating the efforts of all parties required to obtain approval for the Project, including but not limited to the following: Architectural, Engineering, Soils Testing, Surveying, Municipal, Provincial and Federal regulatory authorities.

___ ___ **E)** Consulting in House Design and Planning

___ ___ **F)** Coordination and preparation of all drawings and copies required for mortgage Application, Building Permit, Costing and Construction.

___ ___ **G)** Formation of a Plot Plan.

___ ___ **H)** Coordination and preparation of detailed and itemized construction cost estimates of the entire project, based on the project drawings.

___ ___ **I)** Ensuring competitive trade quotations for construction of the Project are received and reviewing the quotations and pricing received, with a view to recommending to the Owners which trades and suppliers to employ.

___ ___ **J)** Coordinating the preparation of Contracts and Subcontracts as required and approved by the Owners which shall be entered into between the Owners and the Contractor or tradesmen directly.

___ ___ **K)** Arranging and/or applying for all permits, approvals and authorizations as are required from all regulatory authorities.

___ ___ **L)** Assistance in completion of Documents for obtaining Mortgage and Interim Financing for the Project.

___ ___ **M)** Consulting and arranging for execution of all documentation as may be required for the Project including but not limited to the following: Development Agreements, Designing, Engineering and Construction Agreements, Architectural Control Standards, Grade Slip and Survey Certificate or Real Property Report.

___ ___ **N)** Obtaining all insurance required by the Owners for the activities of the Project but not limited to Worker's Compensation Insurance.

___ ___ **O)** Reporting, Delivering and Coordination of all payments to subtrades and suppliers.

___ ___ **P)** Preparing and maintaining all records, financial and otherwise, related to the Project.

___ ___ **Q)** Administering the Owner's construction bank account, and reconciling of same for the Owners' purpose.

___ ___ **R)** Preparing cheques for the Owners' signature(s).

___ ___ **S)** Providing Secretarial and Accounting services as required.

___ ___ **T)** Preparing on a weekly basis, Project status reports, to the Owner.

___ ___ **U)** Coordinating and Formation of a schedule of construction and presentation of schedule to Owners.

___ ___ **V)** Obtaining and coordinating of all city utility and mortgage inspections until completion of Project.

___ ___ **W)** The Consultant shall and does hereby agree to perform the Services in a proper, efficient and workmanlike manner and with that degree of care and skill commensurate with industry standards for competent development coordination and management.

ARTICLE 11 PROJECT MANAGEMENT (CONSULTANT) FEE

2.01 The Owners severally, in accordance with the Proportionate Interests ascribed to them in the Agreement, agree to pay to the Consultant the following fee for the Services:

____

____

____

____

2.02 The Consultant will progress bill for payment of the Services at the end of each payment period. No holdback will be taken on the fee payable to the Consultant.

2.03 All architects, accountants, lawyers, contractors, and other third parties engaged by the Consultant with the consent of the Owners (but not including its other employees) in connection with its performance of these Services will be for the account of the Owners.

ARTICLE 111 TERMINATION

3.01 This Agreement may be terminated by the Owners or the Consultant giving thirty (30) days written notice to the other Party.

3.02 The Consultant shall be paid only for its actual expenses incurred and the fee for its services performed commensurate with work on the Project to the date of termination.

3.03 In the event of termination as described above, the Owners shall have the right to employ forthwith other professionals or individuals to complete the services described herein.

3.04 This agreement will terminate after substantial completion of the project determined by work in progress of the final finishing trades or occupancy of the owners, whichever is first.

ARTICLE 1V ABANDONMENT

4.01 In the event that the Owners wish to abandon the Project for whatever reason, it shall give seven (7) days written notice to that effect and within thirty (30) days after the giving of notice the Owners shall pay to the Consultant its actual expenses incurred and the fee for its services performed to the date of abandonment, based upon the completion of the Project as of the effective date on which the Consultant last provided Services for the Project.

ARTICLE V INDEMNIFICATION

5.01 The Project Manager hereby agrees to indemnify and save harmless the Owners from any and all claims, demands, loss, cost, awards, judgments, actions and proceedings by whom so ever made, brought or prosecuted in respect of loss or damage to or destruction of property or personal injuries, and from any and all loss of, damage to or destruction of property and expense and costs suffered or incurred by the Owners arising out of the negligence of or any wilful act or omission of the Consultant its servants, agents or employees in respect of the Services performed by the Consultant hereunder.

5.02 The Owners proportionate to their Proportionate Interests hereby severally indemnify and agree to hold harmless the Consultant against any claim of or liability to any third party resulting from any action or omission of the Consultant or its agents and employees in conducting operations for the account of the Owners as herein provided; that the Consultant shall not be indemnified or held harmless by the Owners for any loss, damage, claim or liability incurred or created by the Consultant resulting from the negligence or wilful misconduct of the Consultant or its agents and employees.

ARTICLE V1 NOTICE

6.01 Any notice, communication or request to be given to either party shall be in writing by registered mail postage prepaid or by personal delivery addressed to such party at the following address:

As to the Owners:

____

____

As to the Consultant:

____

____

ARTICLE V11 GENERAL

7.01 This Agreement replaces and merges any previous agreements between the parties and comprises the entire Agreement between them. It is hereby declared that there is no condition precedent or warranty of any nature whatsoever, except as stated in this Agreement, and no warranty, condition or covenant whatsoever collateral to this Agreement.

7.02 This Agreement may not be modified or amended except by instrument in writing signed by both parties.

7.03 The parties hereto covenant and agree to do such things and to execute such further and other documents, agreements and assurances as may be necessary or advisable from time to time in order to carry out the terms, conditions and intent hereof.

7.04 If any provision of this Agreement or the application of such provision to any party or circumstance shall be held invalid, the remainder of this Agreement, or the application of such provision to any party or circumstance other than those to which it is held invalid shall not be affected thereby.

7.05 This Agreement shall be interpreted and governed in accordance with the Laws of ______________________

Province or State

7.06 The Consultant may not assign this Agreement without the prior consent of the Owners.

7.07 This Agreement shall enure to the benefit of and be binding upon the parties hereto and their respective heirs, administrators, executors, successors and assigns as the case may be.

7.08 Time shall be of the essence in this Agreement.

IN WITNESS WHEREOF the Parties hereto have executed these presents under their hands and seals or corporate seals as the case may be on the date first above written.

____

Name of Project Manager (Consultant)

____

____

Signed, Sealed and Delivered
in the presence of:

____

Witness

OWNERS

______________________________ ______________________________

Witness

____